Solid Waste Education Recycling

Directory

Teresa B. Jones
Edward J. Calabrese
Charles E. Gilbert
Alvin E. Winder

Northeast Regional Environmental Public Health Center
School of Public Health
University of Massachusetts

 CRC Press
Taylor & Francis Group
Boca Raton London New York

CRC Press is an imprint of the
Taylor & Francis Group, an **informa** business

CRC Press
Taylor & Francis Group
6000 Broken Sound Parkway NW, Suite 300
Boca Raton, FL 33487-2742

First issued in paperback 2020

© 1990 by Taylor & Francis Group, LLC
CRC Press is an imprint of Taylor & Francis Group, an Informa business

No claim to original U.S. Government works

ISBN-13: 978-0-87371-359-7 (pbk)
ISBN-13: 978-1-138-47439-0 (hbk)

Visit the Taylor & Francis Web site at
http://www.taylorandfrancis.com

and the CRC Press Web site at
http://www.crcpress.com

Library of Congress Cataloging-in-Publication Data

Solid waste education recycling directory / Teresa Jones [et al.].
    p. m.
    Includes bibliographical references and index.
    ISBN 0-87371-359-1
    1. Refuse and refuse disposal--Study and teaching--United States.
    2. Recycling (Waste, etc.)--Study and teaching--United States.
    3. Teaching--Aids and devices. I. Jones, Teresa.
    TD793.25.S64   1990
    363.72'8'071--dc20   90-40968
                                                              CIP

Library of Congress Card Number 90-40968

Preface

United States citizens dispose 133 million tons of solid waste each year. Unfortunately we bury 80 to 90% of our refuse, recycle 5%, and incinerate 5 to 10%. During the quest for rational environmental waste disposal techniques, concern and debate have focused on the health issues associated with present and proposed waste management practices. Landfills may contaminate water supplies, incinerators emit acidic gases, organic contaminants such as dioxins, and heavy metals such as lead. Even waste management practices considered environmentally acceptable such as composting and recycling are not without their occupational and environmental risks.

The nation is in the midst of a solid waste crisis demanding our immediate attention. Solutions to our refuse emergency will occur only through creative ideas and effective execution of these ideas. Success in solid waste management practices that protect the public health and the environment depends upon effective public education and participation. Elementary education provides a unique opportunity to offer waste information to children that will enable them to start a life-long learning process leading toward responsible solid waste practices. Secondary school and high school curricula will provide students with the scientific knowledge base to enhance their capacity to make informed waste management decisions. Comprehensive education programs stimulate open discussion between students, teachers, and families, thereby contributing to an increase in a much needed community understanding of waste management problems.

It is the endemic waste management crisis and the importance of waste management education that has prompted the authors to compile a comprehensive list of waste management curricula available in the U.S. The research project started with a data base search to identify existing programs. State and Federal government officials, environmental organizations, academic institutions, and private curriculum development firms were contacted to obtain descriptions and evaluation of their particular programs. A short description and evaluation was developed for each curriculum and these were compiled into a bibliography.

The text summarizes waste management curricula for each state according to categories and features specific to each program. Areas discussed in the program summaries include: the availability of materials to in-state and out-of-state residents; whether permission for duplication is required; the existence of state regulations concerning mandatory recycling; and the existence of state regulations on bottle deposits. The educational infrastructure is discussed with respect to specific curriculum requirements and any requirement of classroom implementation of the program. A project history and description is presented for each program including its status as a state sponsored program, involvement of the Department of Education, the Department of Environmental Protection, and other Departments, and cooperation between departments. In addition, the involvement of environmental groups and academic institutions in program development is noted. Other questions addressed by the states during curriculum development include the following: were other state education materials evaluated, were teachers involved in development, was pilot field testing incorporated into development, is a formal continuing review process of the materials included, and are future revisions to the curricula planned. The distribution of the program is also discussed; this includes the availability of teacher training and in-service credit offerings. The curriculum description section describes whether the program has been validated by the Department of Education, topics in the program, the target age group for the program, the availability of different programs for different education levels (i.e., kindergarten, primary grades, secondary grades, and high school), the inclusion of a teachers guide, and the inclusion of student projects and evaluations. The book also describes those programs containing an outline for a model school waste management program, and references to additional curriculum materials not reported in this text that may be available.

Theresa B. Jones
Edward J. Calabrese
Charles E. Gilbert
Alvin E. Winder

August 1990
Amherst MA

Teresa B. Jones is a research assistant at the Northeast Regional Environmental Public Health Center at the University of Massachusetts. She received a double-major B.A. in History and Environmental Studies from Yale University. As an undergraduate, she directed and expanded a campus residential recycling program and student educational efforts. In her work for the Northeast Center, she designed and conducted primary research concerning national solid waste management strategies and state public education curricula. She then synthesized these data into a comprehensive bibliography of recycling education programs, with suggestions for effective and efficient program development and implementation. Her independent research efforts have addressed the relationship between effective environmental policy and educational programs. Her academic work has focused on the historical, environmental, and economic components of sustainable agriculture.

Edward J. Calabrese is a board-certified toxicologist who is professor of toxicology at the University of Massachusetts School of Public Health, Amherst. Dr. Calabrese has researched extensively in the area of host factors affecting susceptibility to pollutants. He is the author of more than 240 papers in refereed journals and 10 books, including *Principles of Animal Extrapolation, Nutrition and Environmental Health Vols. I and II, Ecogenetics*, and others. He has been a member of the U.S. National Academy of Sciences and NATO Countries Safe Drinking Water committees and, most recently, has been appointed to the Board of Scientific Counselors for the Agency for Toxic Substances and Disease Registry (ATSDR).

Dr. Calabrese was instrumental in the conceptualization and development of the Northeast Regional Environmental Public Health Center and was appointed its first director. The center's mission includes communication, education, and research.

Charles E. Gilbert is a research associate in toxicology at the University of Massachusetts School of Public Health, Amherst. Dr. Gilbert received his BS, MSc, and PhD degrees from the University of Massachusetts.

He was the assistant director for the Childhood Lead Poisoning Prevention Program, Massachusetts Department of Public Health, where he directed improvements in environmental management, case management, and education programs. His research interests are in the area of factors that affect human susceptibility to biological, chemical, and physical agents and how these affect health.

Charles Gilbert worked in the development of the Northeast Regional Environmental Public Health Center and was appointed its first assistant director. This center is a cooperative organization of the New England Public Health Departments and the University of Massachusetts School of Public Health.

Alvin Winder is a Professor of Health Education at the University of Massachusetts School of Public Health, Amherst. He was awarded a B.A. degree in Psychology from Brooklyn College, an M.S. degree in Psychology from the University of Illinois, and a Ph.D. degree in Human Development from the University of Chicago, as well as an M.P.H. degree in Health Education from the University of California at Berkeley.

Dr. Winder has been on the Public Health faculty of the University of Massachusetts since 1978 and a faculty member of the Northeast Regional Environmental Public Health Center since its inception. He has edited and written several books; a text on health education, on adolescence and on group process in education and mental health. He has also authored a number of articles concerned with educating the public about environmental issues. He has additionally produced several video films dealing with environmental pollution. His recent research has focused on educational efforts toward smoking reduction.

Acknowledgment

This project was funded in part by grants from the Center for Environmental Health, University of Connecticut and the U.S. Environmental Protection Agency, Solid Waste Division.

We thank Paula Goodhind and Sandra Orzsulak for their patience, tact, and effectiveness in sustaining the operation of the Northeast Regional Environmental Public Health Center.

This book is dedicated to our families

with love

Contents

INTRODUCTION

Americans produce 133 million tons of solid waste each year, according to a 1986 EPA study. (Haley, 1988) Municipalities bury between 80 and 90% of this waste in "sanitary landfills," recycling approximately 5% and incinerating the remaining 5 to 10%. (Hohman, 1988) The search for environmentally sound waste disposal methods and alternatives has focused scientific scrutiny and debate upon the health issues surrounding present techniques. Many landfills threaten ground and surface water with leachate, residential basements with methane gas migrations, and local atmospheric conditions with organic compound emission. Municipal incinerators emit acid gasses, vaporized heavy metals (e.g., lead), and hazardous organic compounds (e.g., dioxins). Treatment and disposal of residual bottom and fly ash pose additional environmental and public health problems. Recycling and composting depend heavily upon voluntary citizen participation (e.g., source separation) and some processes create toxic by-products (e.g., newspaper deinking).

Despite recent technological improvements in waste disposal techniques, such as leachate collection and water monitoring systems, double liners, electrostatic precipitators, baghouses and scrubbers, new facilities face increasingly stringent state and federal regulations, struggle with economic constraints, and encounter strong public opposition as citizens become aware of selective health issues. Waste stream reduction, reuse, composting and recycling currently offer environmentally sound waste management options. International example provides a model of successful waste management programs based upon this approach. The Japanese recycle approximately 50% of their waste stream, and incinerate two thirds of the remaining, source-separated (and therefore less toxic) materials. The Japanese program emphasizes the social aspects of waste management more strongly than the technical ones. (Nosenchuck, 1988)

The current crisis situation in municipal waste management demands immediate, well-planned, innovative solutions. Significant attention from the political, public and academic sectors indicates increasing widespread concern. Municipalities are presently pursuing a variety of waste management plans—building incincerators or resource recovery facilities, Intermediate Processing Centers (IPCs)/Materials Reprocessing Facilities (MRFs), new "sanitary landfills", organizing collection of recyclables, legislating mandatory composting and recycling laws. The success of recycling and waste reduction programs, as well as intelligent long-term waste management planning, depends upon effective citizen education and participation. Elementary education provides a unique opportunity to encourage developing generations of children to reevaluate their understanding of "waste", examine their own participation in the waste stream cycle, and develop values and skills that enable them to make sound waste management decisions. Secondary school curricula enable students to expand their understanding of scientific and social disciplines to address the more complex and urgent waste management issues. Comprehensive education programs can initiate constructive communication between students, teachers, families, and the community in the interest of expanding recycling and waste reduction efforts.

State environmental protection agencies carry primary responsibility for the formulation and implementation of solid waste management plans. As these strategies begin to require that citizens significantly change disposal behavior, education becomes an essential component of an effective and environmentally sound waste management program. The recycling education research project is an effort to assess state environmental education efforts, especially in relation to technical responses to the solid waste crisis.

This document describes recycling curriculum materials within the context of program structure. This information includes: curriculum description and goals, project history, legislative initiative and funding sources, educational requirements and infrastructure, development procedures, review process and evaluation, field testing and pilot programs, distribution and teacher training, present program status and copyright policy. These data provide sufficient basis for program evaluation and analysis of national recycling education trends. Identification of exemplary efforts forms a central component of broader recommendations concerning implementation of effective education programs. This document should serve to facilitate communication and cooperation between state agencies in the area of environmental education, in the interest of successful solid waste management programs and efficient use of limited state education budgets.

Many states have recognized the need for long-term recycling education programs, and have developed environmental curricula for the public school system. These materials are state-specific to some extent, designed to address the critical waste management issues of a region, to incorporate local environmental priorities, and to accommodate particular educational requirements. Existing materials are, however, readily adaptable to other states' needs, as evident in the repeated implementation of certain materials in many different states. Most states begin curriculum development with a regional or national search for existing materials. Although some agencies compile thorough bibliographies of the materials they collect, limitations on staff and funding force an emphasis on production and distribution of materials. Searches thus tend to be incomplete, and to repeat efforts made previously or concurrently by other states.

Inadequate communication and a lack of overarching organization cause much unnecessary repetition in environmental education efforts. (Pemberton, 1988) Extensive research and materials development has already occurred in environmental education centers, state agencies, nature centers, schools, and national environmental organizations. Pemberton, of the University of Maryland, assesses the status of environmental education: ''The expertise and materials exist now for the development of a national, inclusive, and cost-effective program for environmental education.'' (Pemberton, 1988, p. 11) The Alliance for Environmental Education is currently creating a National Network for Environmental Education (NNEE) to address this administrative problem. A network between regionally based centers will function for research, distribution, teacher training, program and project coordination, evaluation and continual communication concerning environmental education. These centers will have sufficient resources to adequately survey curriculum materials in use within the service region, and make them available to other center-members. The National Network will work with the Educational Resource Information Center (ERIC) to avoid information overlap. The Northeast Regional Environmental Public Health Center's Waste Management Education Research Project specifically addresses some of the problems of inefficient curriculum development and insufficient inter- and intrastate agency communication.

During the 1990s, significant legislation on recycling, ground and surface water protection, air pollution and other environmental issues will increase the need for cooperation on research and educational efforts throughout the country. Enough states have developed effective materials and created comprehensive programs to provide models and materials for national use. A resource base and assessment of these programs is prerequisite to productive transfer of materials between states.

METHODOLOGY

The research project began with a thorough data base search using the Educational Resources Information Center (ERIC) system. States with strong existing programs were identified, and environmental education contact persons recorded. Educational journals, particularly the *Journal of Environmental Education*, were used to identify current trends in education and locate any exceptional activities in the field. During this time period, phone contacts were made through existing waste management and education networks and individual letters sent to programs of particular interest that were located through the data base.

A standard letter describing the research project and requesting information, a contact person and phone number was prepared to conduct the national search. This letter was sent to the Commissioner or Superintendent of each state's Department of Education, to the Director of the Solid Waste Management Division, to the head of the Department of Environmental Protection if the Solid Waste Division was independent of that Department, and to the environmental education contacts provided by the data base search.

A written questionnaire was designed, and translated into an oral interview to be conducted over the phone with the contacts provided by the state agencies. The questionnaire format provided basic guidelines to address the important aspects of curriculum development and implementation. State programs differed greatly and thus some questions were not applicable. The questionnaire was administered in a telephone interview with all individuals referred by state agencies. This process included private curriculum development firms, academic institutions, environmental organizations, and individuals within state government. The focus was primarily upon state-sponsored educational programs; other materials were included if state agencies provided references to specific contacts, or if certain activities appeared repeatedly in materials used by state agencies. Some states did not respond to the survey letter. Of the nonresponding states, only those with educational programs identified by earlier, extensive research were actively pursued for further information. In total, agencies in 33 states were interviewed, 20 of which offered some recycling or solid waste related educational materials.

Many states provided solid waste management information and curriculum materials with return letters, supplementing information provided by the phone interview. All received information was synthesized and written into a draft form for the document. These drafts were returned to the contact person for verification and addition of updated information. Included with the return mailing was a brief project description and a stamped addressed envelope to facilitate validation. Drafts not returned within 3 weeks were assumed accurate, a condition stipulated in the accompanying letter. Verified drafts were then revised, edited, and compiled into the final document text. Some drafts required an additional follow-up call, particularly for programs under development during the original research period. All follow-ups were succeeded with an additional mailing of revised drafts for verification.

REFERENCES

Haley, L. W., Government Refuse, Collection and Disposal Association, "Municipal Solid Waste: The National Scene," Conference on Municipal Solid Waste Disposal: Landfilling, Incineration/Resource Recovery Recycling and Composting, Public Health, Environmental, Economic and Technological Aspects. Amherst, MA April 12, 1988.

Hohman, M., U.S. Environmental Protection Agency, REGION I, "Land Disposal: A Changing State of the Art," Conference on Municipal Solid Waste Disposal: Landfilling, Incineration/Resource Recovery Recycling, and Composting, Amherst, MA April 12, 1988.

Nosenchuck, N. H., Director, Division of Solid Waste, New York State Department of Conservation, "Integrated Solid Waste Management," Conference on Municipal Solid Waste Disposal: Landfilling, Incineration/Resource Recovery Recycling and Composting, Amherst, MA, April 12, 1988.

Pemberton, D. A., "Strategy for a National Network for Environmental Education," *Journal of Environmental Education*, Vol. 19, No. 4, Summer 1988.

SUMMARY OF STATE PROGRAMS IN SOLID WASTE AND RECYCLING EDUCATION

Table 1 provides a comprehensive summary of data compiled on recycling educational programs throughout the country. Only programs sponsored by state agencies are referenced, although private efforts may appear in the document text. Several departments within a state may be represented in one column.

Special table notations:
$ Fee charged for curriculum materials
R Workshop required to obtain curriculum materials
* Special case; see text for specific information

Table Categories:
Materials available in-state
Materials available out-of-state
Duplication permission required

Solid Waste Management
Mandatory recycling
Deposit legislation/bottle bill
Educational Infrastructure
Specific curriculum requirements
Voluntary classroom implementation
Project History and Description
State-sponsored educational program
Department of Education involvement
Department of Environmental
Protection involvement
Other department(s)' involvement
Interdepartmental cooperation
Curriculum Development and Review
Private/professional education developers
Environmental organizations
State university system utilized
Search of other states' materials
Direct teacher input
Pilot project/field test
Formal review/revision structure
Current and/or future revision
Distribution
Teacher training workshop offered
In-service credit offered

Curriculum Description
Validated by the Department of
Education
Recycling/Solid Waste Topics
Other topics
Multidisciplinary
Kindergarten
Grades 1-3
Grades 4-6
Grades 7-9 and/or 10-12
Curriculum package
Teachers Guide included
Student activities included
Student evaluation included
Specific time required
School recycling program outlined
Additional resources available
Non-state sources of materials

Table 1. State Programs in Solid Waste and Recycling Education

	CA	CT	FL	IL	IN	KY	ME	MA	MI	MN	MO	NJ	NY	OR	RI	TX	VT	VA	WA	WI
Materials available in-state	$	X	X	X	X	X	X	X	X	X	X	X	X	X	X	X	X	X	X	X
Materials available out-of-state	$	X	X	X		$			X	X		X		$	$	$		*	$	$
Duplication permission required	X	X			X	X			X		X	X	X	X	X	X	X	X	X	X
SOLID WASTE MANAGEMENT																				
Mandatory recycling	X	X	*									X	*	*	X					
Deposit legislation/bottle bill	X	X					X	X	X				X	X			X			
EDUCATIONAL INFRASTRUCTURE																				
Specific curriculum requirements	*			X	X	X			X	X	X		X		X	X				X
Voluntary classroom implementation	X	X	X	X	X	X	X	X	X	X	X	X	X	X	X	X	X	X	X	X
PROJECT HISTORY AND DESCRIPTION																				
State sponsored educational programs	X	X	X	*	*	X	X	X	X	X	X	X	X	X	X	X	X	X	X	X
Department of Education involvement	X	X	X	X	X	X	*	X	X	X	X	X	X	X	X	X	X	X	X	X
Department of Environmental Protection involvement	X		X	X	X	X	X	X	X	X	X	X	X	X	X	X	X	X	X	X
Other departments involvement	X			X	X	X				X		X						X	X	X
Interdepartmental cooperation	X		X	X	X	X	X	X	X	X	X	X	X	X	X	X	X	*	X	X
CURRICULUM DEVELOPMENT AND REVIEW																				
Private/professional education developers	X					X			X	X			X	X	X	X	X	X		
Environmental organizations							X	X	X	X		X		X	X	X	X		*	
State University system utilized			X				X						X	X	X	X				
Search of other states materials	X			X		X		X	X	X	X	*	X	X	X	X	X	X	X	*
Direct teacher input	X	X	X		*	X	X		X	X	X	X	X	X	X	X	X	X	X	
Pilot project/field test	X	X	X			X	X		X	X		X	X	X		X	X	X	X	X

Formal review—revision structure	X	X										X		X				X
Current and/or future review	X	X		X	X		X	X	X		X	X	X	X	X	X	X	X
DISTRIBUTION																		
Teacher training workshop offered	R	X	*	X	R	X	*	X	X	X	X	R	X	X	X	R	X	X
In-service credit offered		X		X	X		X				X		X			X		X
CURRICULUM DESCRIPTION																		
Validated by Department of Education	X	X	*	X	X		X	*	X	X	X	*	X	X	X	X	X	X
Recycling/solid waste topics	X	X	X	X	X	X	X	X	X	X	X	X	X	X	X	X	X	X
Other topics	X	X	X	X	X	X	X	X	X	X	X	X	X	X	X	X	X	X
Multidisciplinary	X	X	X	X	X	X	X	X	X	X	X	X	X	X	X	X	X	X
Kindergarten	X	X	X	X	X	X	X	X	X	X	X	X	X	X	X	X	X	X
Grades 1-3	X	X	X	X	X	X	X	X	X	X	X	X	X	X	X	X	X	X
Grades 4-6	X	X	X	X	X	X	X	X	X	X	X	X	X	X	X	X	X	X
Grades 7-9 and/or 10-12	*	X	*	X	X	X	X	X	X	X	X	X	X	X	X	X	X	X
Curriculum package	X	X	X	X	X	X	X	X	X	X	X	X	X	X	X	X	X	X
Teachers Guide included	X	X	X	X	X	X	X	X	X	X	X	X	X	X	X	X	X	X
Student activities included	X	*	X	X	*	X	X	X	X	X	X	X	X	X	X	X	X	X
Student evaluation included	X	X	X	X	X	X		X	X	X	X	X	X	X		X		X
Specific time required	X	X		X											X			
School recycling program outlined	X	X		X	X	X			X		X	X			X		X	
Additional resources available	X	X	X	X	X	X	X	X	X	X	X	X	X	*	X	X	X	X
Non-state sources of materials	X	X		X	X	X	X	X	X	X	X	X	X	X	X	X	X	X

California
Solid Waste Management Board
1020 Ninth Street, Suite 300
Sacramento, CA   95814
(916) 322-3330
Contact: Michelle Lawrence
(916) 322-2684
Christopher Peck
(916) 322-8745

SWRL Educational Research and Development
8388 Vickar's St., Suite 117
San Diego, CA   92111
Contact: Roger Scott
(619) 573-1716

Curriculum/Materials:	*The Wizard of Waste* (2-4)
	Trash Monster (5-7)
	Solid waste, recycling, resource conservation topics
	Teachers Guide and curriculum materials
Available:	California Department of Education
	P.O. Box 271
	Sacramento, CA 95802
Fee:	*Wizard*—$20; $2 for *Inservice Guide*
	Monster—$23; $2 for *Inservice Guide*
	Add postage and handling costs
Validated:	Yes
Policy:	California Solid Waste Management Board copyright; Make requests for adaptation rights

Solid Waste Background

California has mandatory statewide recycling and deposit legislation. The Solid Waste Management Board has conducted ambitious publicity campaigns to educate citizens about solid waste issues and recycling.

Educational Infrastructure

The Department of Education produces curriculum guidelines and offers many educational publications for teacher use, but school districts autonomously decide what materials will fulfill state requirements. The Educational Code requires instruction in natural resources and environmental protection topics.

Project History and Description

In 1979, the Solid Waste Board recognized the need for educational materials on solid waste topics. The board initially allocated $100,000 to develop the Solid

Waste Environmental Education Program (SWEEP) through SWRL Educational Research and Development, An additional $100,000 was allocated over 1980-82 to promote and distribute the materials. The Solid Waste Management Board works cooperatively with the Department of Education, communicating through a key contact person.

Curriculum Description

Both curricula are offered for voluntary integration into the classroom. Each Environmental Education kit contains a Teacher's Guide, filmstrip, cassette tape, class poster, picture cards, program records sheet, and 36 of the following pupil materials: booklets, post-tests, self-surveys, badges (stickers), and home information leaflets. Replacements of the student materials are available at a reduced price.

The Teacher's Guide describes the *Wizard of Waste* program as a 2-week unit designed to introduce primary-level children to basic concepts and understanding related to solid waste management, and to help them develop resource conservation skills to use and apply on a daily basis. The program is intended for use in second- or third-grade classrooms, but it is suggested that one grade level be selected to avoid presentation to the same pupils in succeeding school years. Instruction is geared toward three specific outcomes: understanding solid waste concepts, identifying recyclable solid waste, and analyzing and improving personal habits. (The third outcome is not assessed.) The program is organized into ten lessons, each of which can be completed in about 30 minutes. The Teacher's Guide provides background information on solid waste and a glossary of terms.

The *Trash Monster* is a program developed for older students, similar in structure and addressing the main topics of solid waste concepts, solid waste problems, and solid waste solutions. Waste management solutions focus on the four "R's": reverse buying habits, reuse products, recycle solid waste and recover energy. The Teacher's Guide includes procedures for setting up a school recycling campaign.

Curriculum Development and Review

SWRL Educational Research and Development has many professionals trained in the different aspects of curriculum development. SWRL also utilizes consultants with strong education backgrounds, such as environmental educators, for input and suggestions about activities. The group conducted research on curriculum materials available throughout the country, compiling and later selectively integrating what proved appropriate and useful. The development lasted approximately one year (1979), after which time the curriculum was subjected to rigorous field testing and review.

The testing procedures followed an Instructional Systems Design Approach, setting specific desired outcomes for students who participate in the program. A task-content analysis was used to develop prototypes with the appropriate content for each task. "Focus groups" of students in a few pilot schools were tested to evaluate the "learner outcome." The pilot schools served to obtain input from students, teachers, administrators and parents, whose suggestions were incorporated into revisions. The final curriculum was completed in 1980. The Department of Education validated the curriculum materials.

Distribution and Teacher Training

The California Solid Waste Management Board funded the initial 1980 distribution. Through local officials, the Board contacted all school districts with information about the curriculum. Working with the Department of Education, the Board sponsored workshops throughout the state, which were prerequisite to receiving a free curriculum copy. As of July 1982, the Department of Education began to distribute the materials with other educational publications, assuming responsibility for storage and distribution costs. At this time, the present fee schedule was established.

Testing and Evaluation

The curriculum contains pre- and post-tests, student self-surveys and learner outcomes, to allow in-class evaluation. (Scores averaged from 50% on pretest to over 80% on post-tests.) Field testing and SWRL staff observations of classroom instruction were performed as explained in "Curriculum Development and Review."

Current Status

The Solid Waste Board is currently considering a formal review of the curriculum materials. An informal review by the original developer, however, indicated that the materials are still very relevant and, as they are not California-specific, many other states can easily integrate the curricula into their own waste education programs. The Board hopes to allocate approximately $250,000 to update and reprint materials, replacing filmstrips with video cassettes, for example. This money will also be used to reinvigorate the publicity and distribution campaign, which has been less ambitious than in the early 1980s. New efforts will be focused on a few target areas in the state, then expanded.

Revised materials will be available out-of-state for prices similar to those presently listed. Currently, some copies are distributed free to resource individuals, such as librarians, and graduate students.

Additional Information

Recycling materials under development by SWRL—The California Department of Conservation is sponsoring SWRL to develop two educational programs on recycling, for grades K-3 and 4-6. The programs are geared toward increasing recycling behavior in students, teachers and parents. Development began in June 1988, and will be completed for school use in the fall of 1989. The structure will be similar to the *Wizard of Waste* program. The Department of Conservation will fund printing to distribute the materials to 170 schools. Currently no out-of-state policy exists, but similar fees are probable.

Department of Conservation
1416 Ninth St., Rm 1320
Sacramento, CA   95814
(916) 323-4636

REFERENCES

Lawrence, M., California Solid Waste Management Board, Sacramento, personal communication, August 18, 1988.

Peck, C., Communications Advisor, California Waste Management Board, Sacramento, letter, August 17, 1988.

Scott, R., SWRL, San Diego, personal communication, August 18, 1988.

Smith, J. R., Deputy Superintendent, Curriculum and Instructional Leadership, California Department of Education, Sacramento, letter, August 19, 1988.

Trash Monster, California Solid Waste Management Board, 1980.

Waste Education Roundtable, *Final Report: Appendices*, Minnesota Solid Waste Board, August 1986, p. 53.

Wizard of Waste, California Solid Waste Management Board, 1980.

Golden Empire Health Planning Center
2100 21st Street
Sacramento, California

Curriculum/Materials: *Toxics in My Home? You Bet!* (K-12)
"Household Hazardous Wastes," 165 pp., each age group (K-3, 4-6, 7-9, 12-12)
Available: Y
Fee: $6 each without binder; $10 with binder

Each grade grouping is a 1 week course designed to teach students about toxic substances that are commonly found in the home. Specific curriculum concepts and structures vary with each age group. Many states use this curriculum, or adapt it and make the revision available to state teachers. Detailed information concerning each age group curriculum is available through ERIC and from the Golden Empire Health Planning Center.

REFERENCES

ERIC, ED266948, *Toxics in My Home? You Bet!* (K-3), 1984.

Connecticut
Department of Environmental Protection
Information and Education Section
165 Capitol Avenue
Hartford, CT   06106
(203) 566-8108
Contact: Kim Marcy, Recycling Education Coordinator
(203) 566-8108

Curriculum/Materials: ''Hazardous Wastes in Cleanfield, CT''
Follow-up recycling activities (K-8)
''Solid Waste in Riverside, CT''
          Available: Y
                 Fee: Free; may change with increased demand

Solid Waste Background

Connecticut has deposit legislation, and passed a mandatory recycling law in 1987, which will require 25% recycling by 1991. Five resource-recovery facilities and one intermediate Processing Center are now on-line as part of the solid waste management program.

Educational Infrastructure

The Board of Education in each district determines curriculum requirements in all subject areas.

Project History and Description

In 1979, a *Curriculum Activity Guide for Environmental Education* (grades 3-4) was developed. The project utilized a curriculum matrix procedure to address how environmental education could serve to fulfill curriculum requirements in different school districts. This procedure often creates significant incentive for schools to integrate the solid and hazardous waste educational materials.

The waste management education program began in 1984 with the development of a larger plan to improve handling of hazardous waste, particularly household hazardous wastes. The program was designed to obtain local input, encouraging participation by school administrators, teachers, town committees and interested organizations in the planning. The program has three components: technical support, curriculum distribution and teacher training, direct service programs with students. The Department of Environmental Protection prefers to offer the student service program in conjunction with hazardous waste cleanup days to reinforce classroom learning and encourage participation. The program offers adult educational materials through the Community Environmental Education Program (CEEP) and Recycling Education Program.

The Department of Environmental Protection worked independently on the project.

Curriculum Description

The "Hazardous Waste in Cleanfield, CT" is part of the direct service program and centers around a role-play performed by students after learning about hazardous wastes. The packet includes background materials for the teacher, problem sheet, and the "Hazardous Substances Simulation Game." The Information and Education Unit conducts in-class lectures and demonstrations, using multi-media presentations, over a 2- to 3-week period, emphasizing "multiple exposure" to important topics. The curriculum is based on the National Wildlife Federation *Conservation Learning Activities for Science and Social Studies Project* (CLASS). The Information and Education Section information sheet explains that the curriculum "involves students in a real life situation of siting a hazardous waste disposal site in their city. They apply knowledge they have learned, assuming the roles of town officials such as mayor, town health official, members of the Conservation Commission, business people, chemical hauling company executives and hazardous waste treatment specialists." Students learn what hazardous wastes are, where they come from, and how they are used in household products. The decision is debated in a town meeting forum.

The curriculum is very flexible to local curriculum objectives and can be adapted by Information and Education staff. It does require homework assignments and a considerable amount of follow-up work by the teacher.

The "Solid Waste in Riverside, CT" activities center around a role-play. The packet provides background information on solid waste, teacher background materials about the role-play, and descriptions of the various actors involved in siting a landfill or trash-to-energy facility.

The Follow-up Recycling Activities teaches students what they throw away every day and which of these materials can be recycled. The activities are all taken from the Washington State *A-Way with Waste* curriculum.

The Information and Education Section offers the Golden Empire Health Planning Center curriculum, *Toxics in My Home? You Bet!* for grades K-12.

Curriculum Development and Review

The curriculum programs branched from the original hazardous waste project. The Information and Education Section received curriculum materials from other states and organizations, although a thorough nationwide search was not conducted. Information and Education staff developed the materials, drawing from other programs, and made the Connecticut-specific service program available in 1984.

The recycling materials were selected from the Washington State *A-Way with Waste* program; the Washington materials were relevant, easily integrated and adaptable to Connecticut's needs.

Distribution and Teacher Training

All materials are made available upon a voluntary basis at no cost. Service programs are offered upon request to schools. In 1987, the Department gave presentations in 10 to 15 school systems. The Solid Waste and Recycling packets are distributed through in-service teacher training workshops conducted by Information and Education staff, and after appearances by "Ray Cycle", Connecticut's recycling superhero. The workshops are designed for teachers of all grade levels and stress a hands-on approach.

Testing and Evaluation

The town meeting forum allows teachers to assess informally how much students have learned about hazardous and solid waste topics. The recycling activities have the pre- and post-tests of the original curriculum.

REFERENCES

Fish, S., personal communication, July 10, 1988, and August 22, 1988.

"Hazardous Waste Education Program," information sheet provided by Connecticut Dept. of Environmental Protection, Hartford, July 1988.

Marcy K., Connecticut Dept. of Environmental Protection Recycling Education, personal communication, June 5, 1989.

Pragman, T., Connecticut Dept. of Environmental Protection, Hartford, personal communication, June 1988.

"The Adventures of Ray Cycle," information sheet, Connecticut Dept. of Environmental Protection, Hartford, 1988.

Florida
Department of Education
Florida Education Building Suite 444
Tallahassee, FL   32399-0400
Contact: Judy Brenggeman
Coordinator, Office of Environmental Education
(904) 487-8414
Contact: Roy I. King, Program Specialist for
Environmental Education
(904) 488-2601

Curriculum/Materials: In progress
Waste Management, Recycling Topics
Waste Recycling Resource Guide and Curriculum
materials
Available: Fall 1990
Fee: Free

Solid Waste Management

In 1988, Florida passed a solid waste management act that included a specific mandate concerning recycling awareness and waste reduction education. The legislation calls for a mandatory disposal ratio of one third landfilling, one third incineration and one third recycling of the waste stream by 1994. The Department of Education, in cooperation with the state university system and the Department of Environmental Regulation, must develop guidelines for recycling and waste reduction education, for student and adult audiences. A study guide for this educational program has been completed and both the study guide and the educational guidelines are available August 1990. The local school boards must ensure that such education programs are provided for elementary and secondary school students.

Educational Infrastructure

The Appropriations Act created two new positions within the Department of Education, in Environmental Education and Recycling Awareness. These two positions are primarily responsible for developing and distributing curriculum materials. Local school boards are autonomous in determining what curriculum materials are used in each district.

Project History and Description

The 1988 Appropriations Act allocated $132,000 from the general state fund to the education program as defined under the Act. The curriculum must be developed in agreement with the Department of Environmental Regulation solid waste management guidelines, and with a technical accuracy as defined by solid waste experts.

Curriculum Description

The legislation requires recycling awareness and waste reduction education. Overall goals of the program are geared toward ultimately changing behavior in regards to recycling activities and solid waste production.

Curriculum Development and Review

The Department of Education will first conduct a survey of all school districts in Florida, to assess what materials they are already using that address recycling and solid waste topics. Depending upon the results of this survey, the Department may adapt, produce and distribute existing materials. The Department is gradually compiling educational materials from other states to selectively utilize in the Florida curriculum program.

The actual development of activities may utilize state university resources, and will be a cooperative venture with teachers from around the state. The Department may hire teachers during the summer to participate in activity development. Materials will be reviewed, and finally edited by the Environmental Education and Recycling Awareness professionals.

Testing and Evaluation

The Department will field test the materials. The extent of testing will depend upon to what degree new materials are introduced into the schools. If adapted existing materials form the core of the curriculum, extensive field testing is not deemed necessary.

REFERENCES

Cullar, W., Deputy Director, Division of Public Schools, personal communication, August 23, 1988.

McDannel, C. A., and King, R. I., Florida Department of Education, November 27, 1989.

Phillips, J., Florida Department of Education, July 17, 1990.

Sherwood, C., Florida Department of Education, personal communication, March 21, 1989.

Van Fleet, D. S., Director, Division of Public Schools, Department of Education, letter, August 10, 1988.

Illinois
Environmental Protection Agency
Office of Public Information
2200 Churchill Road
P.O. Box 19276
Springfield, IL 62794-9276
(217) 782-3397

Curriculum/Materials: *The Land We Depend On* (5)
Land Pollution, Solid and Hazardous Waste Topics
Student activities, teacher recommendations, posters
Available: January 1989
Fee: Free
Validated: N
Policy: Materials may be duplicated for classroom use

Solid Waste Background

Illinois does not have statewide mandatory recycling, although state agencies use recycled paper for printing.

Educational Infrastructure

The Department of Education does not mandate specific curriculum requirements.

Project History and Description

The Illinois Environmental Protection Agency is committed to promoting environmental education. In 1987, the EPA implemented comprehensive environmental awareness and a poster exchange. This program was entitled: *Water: The Liquid of Life* and focused on the importance of clean water. The water project began as an in-house idea for a poster contest. Interaction with the Department of Education encouraged expansion of that idea to an "educational exchange" involving curriculum materials for the fifth grade. The project expanded to include the Divisions of Public Water Supply and Water Pollution Control, which shared funding responsibilities and assisted with technical aspects. The Board of Education consulted on identification of and communication with regional superintendents.

The Agency will sponsor a similar program in January of 1989. The program is entitled: *The Land We Depend On*. This will be available to educators after January 1989.

Curriculum Description

The new program will focus on recycling, landfilling, composting, hazardous waste and incineration. It is designed as a 1-week multidisciplinary program. The format will be similar to that of the water education materials, which included five "modules", each with teacher recommendations, background materials, facts, student activities, experiments and evaluation procedures. Posters accompany educational materials.

Curriculum Development and Review
Public Information Officers, staff from the Divisions of Public Water Supply and Water Pollution Control, and a masters candidate developed the materials. Information from other states was collected and selectively used. For the land pollution materials, the Division of Land Pollution Control will ensure that all technical information is accurate. Final compilation and editing was done by the masters candidate. The Graphics Department is responsible for formatting and production.

Distribution and Teacher Training
The Illinois EPA asked the state's Regional Superintendents to notify all schools about the "educational exchange", providing teachers with the option to participate. Requests for material packets must be received by a certain date.

The IEPA does not offer teacher training workshops; the materials provide detailed teacher instructions.

Testing and Evaluation
Each section contains student evaluation questions.

Additional Information
Since the materials are designed for a 1-week program and do not claim to fulfill curriculum requirements for any particular discipline, the IEPA did not seek validation for the materials.

Current Status
Materials will be available in January 1989.

REFERENCES

Ferguson, G., PIO, IEPA, personal communication, August 23, 1988.

Killian, B. P., Director, Illinois EPA, letter, August 4, 1988.

Muraro, J., Public Information Officer, IEPA, personal communication, August 23, 1988.

Indiana
Office of School Assistance
Center of School Improvement & Performance
Room 229, State House
Indianapolis, IN   46204-2798
Contact: Joe Wright, Environmental Science Consultant
(317) 232-9141

Indiana Department of Commerce
Division of Energy Policy
1 North Capitol, Suite 700
Indianapolis, IN   46204-2288
Contact: Monica Cannaley
(317) 232-8955

Curriculum/Materials:	*Recycling Coloring and Activity Book (K-6)*
Environmental Education	
materials:	*Total Environmental Education* (K-12)
	Whole Earth Design (4-12)
	Hazardous Dumping Grounds Module (6-12)
	Take Pride in America (TPIA) (K-12)
	The Outdoor Classroom (K-6)
Available:	Coloring book available January 1989 from Department of Commerce
	Other resources available through ERIC
Validated:	NA

Solid Waste Background

Recent legislation channels some oil overcharge funds to be utilized for recycling programs. This money will be used for educational and promotional programs. The Department of Environmental Management is responsible for state solid waste strategies. No statewide mandatory recycling legislation exists, although some communities have local resolutions or voluntary programs. The Clean City Committee of Indianapolis is currently trying to implement some recycling suggestions offered by the Recycling Commission, such as supplying mobile recycling trucks and undertaking an organized advertising campaign. Fifty thousand dollars has been allocated for the state Recycling Pilot Program budget.

Educational Infrastructure

Indiana has mandated *Proficiencies* for each instructional area in the schools. Students are tested on these requirements through the Indiana State Testing for Educational Progress (ISTEP) program. Environmental Education goals have been incorporated into science requirements, and partially into social studies guidelines.

The state coordinates efforts for environmental education between several agencies: Department of Commerce, Department of Natural Resources, Department of Public Instruction, and the Board of Health.

Project History and Description

The Department of Public instruction produces curriculum materials and guidelines, which are provided to teachers through in-service programs and workshops throughout the state. The Department of Education encourages local school districts to develop and implement "total curriculum programs" incorporating environmental education goals as set forth in the documents listed above. The Environmental Education materials are not available directly from the Department of Public Instruction, but can be accessed and purchased from the Educational Resources Information Center (ERIC).

The Division of Energy Policy sponsors an annual "Indiana Energy Month" in the Indiana public schools. The promotional budget for this program is $20,000. The program features lessons and activities focusing on one particular topic; the October 1988 Energy month will emphasize recycling. The program features a variety of contests, such as essays, posters, bumper stickers, magnets, and other educational and promotional materials. Materials are judged and selected for some state promotional or educational use. The Division sponsors an award ceremony at the end of the month.

Curriculum Description

The *Coloring and Activities Book* will have creative pictures, word-searches, crossword puzzles and other student activities emphasizing recycling topics.

Curriculum Development

Elementary school students will create pictures and activities that are related to recycling. Student work will be submitted to the Division of Energy Policy for evaluation and synthesis. The best activities will be selected and incorporated into a *Coloring and Activities Book,* which will be printed and distributed throughout the state.

Distribution and Teacher Training

The Division of Energy Policy mails Energy Month information to every school; the *Coloring and Activities Book* can be distributed through the same mechanism.

Testing and Evaluation

Not applicable.

REFERENCES

Cannaley, M., Indiana Division of Energy Policy, Department of Commerce, Indianapolis, personal communication, August 29, 1988.

Total Environmental Education, Indiana Dept. of Public Instruction, Indianapolis, 1978.

Whole Earth Catalog, Indiana Dept. of Public Instruction, Indiana State Board of Health, 1976.

Wright, J. W., Environmental Science Consultant, Department of Education, Indianapolis, personal communication, August 26, 1988.

Kentucky
Center for Math, Science & Environmental Education
Western Kentucky University
Bowling Green, KY   42101
Contact: Joan Martin, Program Coordinator
(502) 745-4424

Curriculum/Materials: *Waste: A Hidden Resource* (7-12)
Solid Waste Management topics
Unit activities, State Curriculum Guide
Available: Y
Fee: Unknown

Project History and Description

Western Kentucky University is one of two Kentucky members of the Tennessee Valley Authority Environmental/Energy Education Program (see Tennessee Valley Authority). Ms. Joan Martin provided the following description of the Center: The Center and the Tennessee Valley Authority (TVA) entered into a 5-year contract to join the TVA-initiated University Centers for Energy and Environmental Education. As a part of this network, the Center meets with representatives from the participating University-based Centers, publishes a quarterly newsletter, and develops workshops to train teachers in the use of materials such as *Project Class, Project Learning Tree* and *Project Wild*. Additionally, the Center participated in the development and field testing of teacher-generated energy and environmental education materials, such as *Waste: A Hidden Resource* and *Energy Sourcebooks: High School Unit, Sixth Grade Unit,* and *Grades 3-5*.

Educational Infrastructure

The Department of Education has centralized authority to mandate specific curriculum requirements. State Curriculum Guides outline these requirements, while individual school districts can determine how best to fulfill them.

Curriculum Description

The activities are designed to supplement waste education in the existing 7-12 curricula in four major subject areas: Science, Social Studies, Language Arts, and Mathematics. Activities are arranged in four chapters: "Overview of Solid Waste," "Hazardous Wastes," "Municipal Waste," "Simulation: Crisis in Center City." The first three chapters begin with fact sheets to provide background information. All curriculum materials are accompanied by the state *Curriculum Guides* for respective disciplines. Each state serviced by TVA must combine its own curriculum guide with *Waste: A Hidden Resource* and incorporate supplemental activities into study plans.

Curriculum Development and Review

Curriculum development began in the summer of 1985. Teachers were recruited from different disciplines, and were paid travel expenses and stipends to participate in activity development. All participants first attended a series of workshops presented by the Tennessee Valley Authority and representatives from regional industry and business. These workshops featured personnel involved with solid waste management, providing information on solid waste problems. Teachers then created instructional activities during two separate development sessions, one in the summer of 1985, and one during the winter of 1985-86. These activities were pooled, reviewed, and edited by the Center for Math, Science & Environmental Education. The Center and the TVA Environmental/Energy Education Program coordinated the development of the technical information into classroom activities. In the spring of 1986, classroom teachers field-tested the activities to ensure appropriate teaching methodology and practices. The Center provided input for final revisions. The curriculum was completed in early 1987.

Distribution and Teacher Training

The Center networks with TVA and the Professional Development Center Network (PDCN). Ms. Joan Martin explained how the Center cooperates with the PDCN to work with the schools: The PCDN is an established consortium consisting of Western Kentucky University (WKU) and 28 rural school districts in Western Kentucky, joined together for in-service training. The CSMEE is a primary source for these training programs, which are adjusted in direct response to annual needs assessment conducted through the PDCN. It is through the monthly meetings of participating instructional supervisors that the Center recruits teachers for development of materials, for environmental education training workshops, and for product distribution workshops.

Western Kentucky University offered the first teacher workshops in the TVA service region. These workshops were also attended by representatives from other TVA University-based Centers, who provided similar workshops to their teachers. Workshop attendance was a prerequisite to receiving a free copy of the supplemental curriculum material. The University-based Centers also trained participating teachers to conduct "back-home" workshops to ten additional teachers, who each received a copy of the materials. Western Kentucky University began this distribution process in 1987, and has distributed approximately 300 copies. Other states in the TVA service region began distribution in early 1988.

Testing and Evaluation

The Tennessee Valley Authority assumed responsibility for curriculum evaluation. A questionnaire was collected from teachers who had implemented the curriculum. The evaluation asked teachers to rate specific activities in four categories: ease of presentation (materials availability, procedural sequence, background information); student interest; appropriateness (subject appropriateness, grade appropriateness, duration); effectiveness (objectives, graphics, student sheets, activity evaluation, extension activities). The Office of Environmental/Energy Education is presently evaluating these questionnaires and making appropriate curriculum revisions. The Center provided a summary of the teacher evaluations which identified weaknesses to be addressed by future revisions, and indicated overall that the activities are "successfully serving the purpose for which they were developed."

Current Status

The Center for Math, Science and Environmental Education continues to offer workshops for teachers interested in integrating environmental issues into their curricula. Further curriculum revision will address the readability of student materials, length of activities, materials availability, and recommended grade appropriateness.

REFERENCES

Judy, John, TVA, Division of Land and Economic Resources, Environmental/Energy Education Program, Project Manager, personal communication, August 3, 1988.

Martin, Joan, Center for Math, Science and Environmental Education, personal communication, August 2, 1988.

Martin, Joan, Program Coordinator, CSMEE, letter, August 13, 1988; and accompanying information ("CSMEE: How We Network to the Schools").

Seppenfield, A., Environmental Education Consultant, Kentucky Department of Education, Frankfort, letter, July 27, 1988.

Waste: A Hidden Resource, Introduction, p. v, and "Summary of Teacher Evaluations."

Kentucky
Natural Resources and Environmental Protection Cabinet
Division of Solid Waste Management
18 Reilly Road
Frankfort, KY   40601
Contact: Chris Sanders
(502) 564-3350

Office of Communications
Capital Plaza Tower, 4th floor
Frankfort, KY   40601
Contact: Jodee Siebert
(502) 564-2672

Curriculum/Materials: *Adventures of Colonel Kentucky* (5-6)
Waste Education topics
Student Workbook, Teachers Guide
Available: Y
Fee: .30 Workbook; .35 Teachers Guide
Limited to one copy
Validated: Y

Solid Waste Background

Each county in Kentucky submits a 20-year solid waste management plan, which is revised every 4 years. The Cabinet approved these plans in 1987. Most plans give some attention to resource-recovery and recycling. Kentucky does not have mandatory local or statewide recycling, but the Cabinet actively promotes recycling activities and is seriously considering recycling legislation for the 1990 legislative session. Kentucky's most severe solid waste problem is illegal open dumping, despite strict sanitary landfill regulations. The state has difficulty obtaining 100% participation in collection, although the services are offered to all residents.

Educational Infrastructure

The Department of Education has centralized authority to mandate specific curriculum requirements. State Curriculum Guides outline these requirements; individual school districts can determine how best to fulfill them. The Natural Resources Cabinet tries to work closely with the Department of Education to ensure that teachers can easily use the materials provided.

Project History and Description

In 1985, the Natural Resources Cabinet decided to begin development of an environmental education program. The Department had no materials for the public schools at that time. After conducting a search of organizations that developed educational materials, the Cabinet contracted with a private educational development firm in California, Innovative Communications. Innovative Communications, in close cooperation with the Cabinet, developed materials concerning water, to encourage students in grades 3-4 to "observe, conserve and preserve" Kentucky's water resources. The recycling packet was developed by the same educational firm.

Curriculum Description

The 16-page *Adventures of Colonel Kentucky* features Kentucky characters relating information about the state, litter, recycling, and solid waste topics, and fun activities for the classroom. The teachers guide provides background information and activity suggestions, offering a conceptual sequence and curriculum outline. The program is flexible, designed to fit into whatever time period is available. The materials include "lesson extenders" which allow teachers to expand the program.

Curriculum Development and Review

The curriculum packet on litter and recycling for grades 5-6 was developed through an identical contract with Innovative Communications in 1987, and completed in 1988. The general development process followed by the private educational development firm begins with extensive consultation with the client. In Kentucky, this included: the Secretary of the Natural Resources Cabinet, several Superintendents of Public Instruction, and Resource Cabinet staff. The consultation included an evaluation of the educational system, including specific curriculum requirements. The two curriculum development professionals also studied the state waste management characteristics. Continual communication between the Cabinet and the educational development firm insured that materials were sensitive to Kentucky-specific needs. The firm generally works with a teacher advisory committee. Original materials are reviewed within the firm, and field testing is performed as part of the contract. Results are either directly incorporated into revisions or a report is submitted with the materials to the client. Innovative Communications will also provide for teacher evaluation, materials production and marketing, depending upon client needs.

The Cabinet paid travel expenses for volunteer teachers to work together for 2 days to revise the recycling and litter materials. The Cabinet offers a supplemental curriculum packet, which includes audiovisual materials, speaker lists, and other resources for interested teachers. The Cabinet will give lectures and workshops on solid waste topics.

Distribution and Teacher Training

The Natural Resources Cabinet communicated with the Instructional Supervisors of each public school, explaining the available materials and the necessary process to obtain them. (This was done originally for the curriculum materials on water resources in 1985-86.) Teachers were required to attend workshops given by the Cabinet, where they received a free copy of curriculum materials. In-service policies differed between counties, but each workshop lasted a minimum of 45 min, and all teachers were awarded credit for attendance. Schools or community organizations could request workshops. The Cabinet sought sponsorship from local environmental organizations (e.g., garden clubs) to help fund the workshop effort. In-service for the litter and recycling program began August 1988.

Testing and Evaluation

All teachers who attended workshops were given a voluntary questionnaire to complete after implementing the curriculum. The Cabinet reports only "positive" responses thus far.

Additional Information

The Natural Resources Cabinet also offers educational materials concerning air and water quality.

Innovative Communications is a private team of designers, illustrators and curriculum specialists. They produce a wide range of multimedia educational and informational materials for industry, government and private groups.

Innovative Communications
207 Coggins Drive
Pleasant Hill, CA   94523
(415) 944-0923
Contact: Bob Johnson, Paul Fletcher

REFERENCES

Hayden, A., Division of Waste Management, personal communication, August 3, 1988.

Siebert, J., Office of Communications, personal communication, August 2, 1988

Wedl, C., Marketing Manager, Innovative Communications, Pleasant Hill, CA, August 4, 1988.

Maine
Office of Waste Recycling and Reduction
State House Station 130
Augusta, ME   04333
Contact: BJ Jones
(207) 289-3154

Curriculum/Materials: Classroom Activities (K-12)
                             Recycling, Solid Waste Topics
          Available: Fall 1989
          Validated: N

Solid Waste Background

The Maine legislature passed a Solid Waste Act in 1987, creating the Office of Waste Recycling and Reduction in April of 1988. The new statewide plan focuses on landfill closure and remediation, and municipal assistance with recycling programs. Education is one major component of the plan, although specific funds were not earmarked for this purpose. Maine does not currently have mandatory recycling, but does have deposit legislation.

Educational Infrastructure

There are no specific curriculum requirements concerning the recycling educational materials. The activities will be voluntarily integrated into appropriate classroom subjects.

Project History and Description

The Office of Waste Recycling and Reduction submitted its first Recycling Plan, which included education measures, in January of 1989. The Office collected other states' educational materials during this time, in preparation of its educational program planned for implementation in the fall of 1989.

Curriculum Description

The curriculum is a collection of multidisciplinary activities selected from existing programs in other states. Some activities will be modified, with permission, to include Maine-specific information.

Curriculum Development and Review

The Office of Waste Recycling and Reduction originally collected materials from other states, and selected a compilation to be distributed to Maine teachers. Teachers provided direct input during a workshop held at the University of Maine during the spring of 1989. The advisory committee consists of representatives from the Department of Education, Science Teachers Association, Environmental Educators Association, Maine Audubon Society, and the Office of Waste Recycling and Reduction. Teacher evaluations and suggestions will be incorporated by the committee into the final development and adaptation.

Distribution and Teacher Training

The original packet has been distributed to volunteer teachers for a preliminary evaluation. The adapted revised materials will be distributed through teacher training workshops which the Office plans to conduct beginning in the fall of 1989.

Testing and Evaluation

Student evaluation measures will depend upon the structure of the original curriculum from which materials were taken. The Office has encouraged direct teacher input concerning implementation of selected materials.

REFERENCES

Bither, E. M., Commissioner, Dept. of Educational and Cultural Services, letter, August 3, 1988.

Jones, B. J., Office of Waste Recycling and Reduction, personal communication, August 18, 1988.

Jones, B. J., personal communication, June 2, 1989.

Massachusetts
Department of Environmental Quality and Engineering
(Department of Environmental Protection, 8-89)
Division of Solid Waste Management
1 Winter Street, 4th floor
Boston, MA   02108
(617) 292-5960
Contact: Linda Domizio
Public Information Officer
(617) 292-5988

Curriculum/Materials: Resource Materials for Schools (K-12)
Solid Waste Management Topics
Available: Y—Fall 1989
Validated: N

Solid Waste Background

The Massachusetts Solid Waste Act was signed in 1987, to assist municipalities with landfill cleanup and expansion, recycling and composting, and new facility construction. Deposit legislation currently encourages return of recyclable beverage containers, although no mandatory recycling legislation exists. A model Materials Recovery Facility (MRF) in Springfield is scheduled to come on line in the fall of 1989. Eighty-five communities will source separate and send recyclables to this facility. The state is encouraging private construction of MRF's and is working actively on market development as part of its integrated solid waste management strategy.

Educational Infrastructure

Local school boards determine specific requirements within the school system. The DEQE materials packet will be voluntarily integrated into appropriate classroom subjects.

Project History and Description

The resource materials for schools are part of a state public education program on recycling and solid waste management. The Department of Environmental Quality and Engineering (DEQE) began a curriculum development project in 1986, which involved compilation of educational materials from other states. The Public Information Officer then selected from these materials, and worked with an advisory committee to prepare more in-depth teacher background information relevant to Massachusetts. Distribution of the material resource packet will occur in conjunction with an ambitious publicity campaign on recycling and solid waste management.

Curriculum Description

The resource materials address many issues of solid waste management, including recycling, landfilling, composting, incineration, waste reduction and hazardous waste. The packet offers interdisciplinary classroom activities, teacher background information, and a list of further resources available from the state. The packet also includes provisions for setting up recycling programs in the schools.

Curriculum Development and Review

After compilation of curriculum materials from other states, the Public Information Officer selected appropriate activities for inclusion in the packet. Working with an advisory committee, she then developed more in-depth teacher background information to accompany the classroom activities. These materials should be available to teachers during the 1989-1990 academic year.

Distribution and Teacher Training

The DEQE hopes to offer teacher training workshops during 1989-1990 to accompany distribution of the materials packet.

Testing and Evaluation

Information not available.

REFERENCES

Domizio, L., Public Information Officer, DEQE, Boston, personal communication, June 1, 1989.

Ellis, S., Regional Planner, DEQE, Springfield, MA, personal communication, August 2, 1988.

Jones, D., Assistant Commissioner of Education, Massachusetts Dept. of Education, personal communication, August 17, 1988.

"Questions and Answers on the Solid Waste Act of 1987," DEQE, DSWM, January 1988.

Roy, N., MA DEQE, DWSM, personal communication, August 17, 1988.

Michigan
Department of Natural Resources
Stevens T. Mason Building
Box 30028
Lansing, MI 48909
Contact: Marta Fisher
(517) 373-0540

Curriculum/Materials: draft stage
Waste Information Series for Education (WISE) (K-12)
Waste Management, Resource Recovery topics
Teacher's Guide, student activities
Available: Y—Fall 1989
Fee: free (if supplies are sufficient)
Policy: Department of Natural Resources copyright

Solid Waste Background

Michigan passed the Clean Michigan Fund Act in 1985, the goal of which was to reduce dependence on current solid waste disposal methods. The Act provided grants to encourage communities to develop and implement resource recovery strategies. Grant projects include recycling and composting programs, hazardous waste collection days, feasibility studies, waste to energy facilities, market development and educational programs. The curriculum described received funding through this legislation. Michigan also has a bottle bill and considerable local recycling legislation.

Educational Infrastructure

The state sets curriculum requirements, but school districts have significant autonomy in determining how to fulfill those requirements. Michigan does not currently have a statewide environmental education program.

Project History and Description

The project began due to initiative within the Department of Natural Resources (DNR). Using money from the Clean Michigan Fund, the DNR hired a consulting firm, Franklin Associates, Ltd., to develop the materials. The Department of Education provided some initial input to the project, and will consult concerning distribution and promotional procedures. Apart from this interaction, the curriculum is an independent DNR project.

Curriculum Description

The program is designed as a week-long curriculum, to be integrated into a variety of instructional areas. It is, however, primarily designed for science classes. The lessons are divided into grade groups—K-3, 4-6, 7-9, 10-12. The core of the curriculum are the "Student Backgrounders", news-type magazines for each grade level containing articles concerning solid waste management. These articles form a basic text around which the teachers can build lessons; the other materials in the curriculum are related to the "Backgrounders". The curriculum package includes teacher background materials, student activities, a glossary, audiovisual materials, computer games, and resource bibliography. Since accurate information about solid waste changes over a short time, the "Backgrounders" will probably require revision within a few years; other curriculum materials would also be updated.

Curriculum Development and Review

Franklin Associates, Ltd., Engineering/Environmental/Management Consultants develops and oversees piloting and evaluation of curriculum materials. The firm works with DNR staff and a teacher advisory committee. Franklin Associates, Ltd., began the development with a nationwide search for other materials in waste education. Subcontractors are used to research and write the "Student Backgrounders", develop video tapes, films strips and computer games to supplement the materials. A secondary education specialist is primarily responsible for the project, who works with a primary educator. They have developed lesson plans, consulting with the Michigan teachers. The firm did not work with any specific curriculum requirements. The project reached draft stage in June 1988.

Distribution and Teacher Training

The Department of Natural Resources plans to distribute a minimum of one copy of the materials to each school in the state, approximately 5,000 schools. Teachers who request separate copies will be supplied with any remaining materials. The DNR plans to print 7,000 copies.

The Department of Natural Resources will probably conduct training workshops, as they consider it a very effective way to distribute materials and encourage teachers to use them. The curriculum will not be available for distribution until the fall of 1989; no finalized plans for teacher training have been established.

Testing and Evaluation

The curriculum will be field tested in the fall of 1988. Revisions using results from this testing will be done during the winter of 1988-89. Evaluation procedures are part of the contract with Franklin Associates, Ltd., and are not finalized at this time

Current Status

The materials will be available in the fall of 1989.

Additional Information
Franklin Associates, Ltd., Engineering/Management Consultants
4121 West 83rd St., Suite 108
Prairie Village
Kansas City, MO   66208
(913) 649-2225
Contacts: Marge Franklin, Bob Hunt

Franklin Associates, Ltd. primarily performs technical consultation of solid waste management projects, and develops some educational materials concerning solid waste topics.

REFERENCES

Fisher, M., DNR, personal communication, August 25, 1988.

Franklin, M., Franklin Associates, Ltd., Kansas City, MO, personal communication, August 25, 1988.

Howard, A., Chief, Waste Management Division, Department of Natural Resources, Stevens T. Mason Building, Box 30028, Lansing, MI, letter, August 16, 1988.

Mincemoyer, N. C., Coordinator, Science Education, Department of Education, Lansing, MI, letter, August 22, 1988.

Michigan
Genesee County Cooperative Extension Service
G-4215 W. Pasadena Ave.
Flint, MI   48504-2376
Contact: Henry Allen
(313) 732-1474

Curriculum/Materials: *Increasing Solid Waste Awareness in the Classroom: Lessons in Resource Recovery* (K-12) Recycling, Reduction, Reuse and Composting Lesson Plans, Students Activities
Available: Y
Fee: Duplication and postage costs
Policy: Department of Natural Resources Copyright Video tapes may be duplicated

Project History and Description

The Genesee Cooperative Extension Service conducted an initial survey of all public schools in the county, to determine the status of waste education. Results indicated that no materials were in use; the Service received a 1-year grant in 1987 to develop waste education materials. The grant provided a budget of approximately $20,000.

Curriculum Description

The materials are a series of 15 activities plus games and puzzles designed to "promote awareness" of solid waste issues. The activities can be used for several grades with minor revisions performed by the teacher, and are multidisciplinary in nature.

Curriculum Development and Review

The Extension Service has a permanent network of teachers, which was tapped to obtain volunteers to work on the curriculum project. These teachers, representing many different disciplines (e.g., Science, Spanish, English, Social Studies) and Agency representatives staff formed an advisory committee. Flint, Michigan has one school that strongly emphasizes environmental education. Ms. Cotner consulted frequently with teachers from this school, and later field tested the curriculum through this cooperation.

The advisory committee originally evaluated other materials, brainstormed ideas, and created tentative activities. Ms. Cotner revised the materials over a 3-month period, and submitted the revision to the committee for review. After approval, the materials were printed and made available to county teachers.

Distribution and Teacher Training

Ms. Cotner sent a letter to all schools in the county, explaining the materials and offering to give workshops. The existing communication network of the Extension Service also functioned to spread information. Ms. Cotner then responded to school requests for workshops, and distributed materials free at these sessions. The Extension Service also developed two video tapes providing background on solid waste topics and the curriculum. These tapes could be used to substitute for a training workshop, and are also useful for general audiences.

Currently, the Extension Service sends to schools order forms and posters to promote the curriculum.

Testing and Evaluation

The curriculum was presented with a demonstration during a teacher in-service day. Suggestions were then incorporated into final revisions by the advisory committee before the curriculum was printed.

Teachers receive a questionnaire about the curriculum, which is returned to the Extension Service. Records are kept of all order forms and teacher suggestions. The Extension Service hopes to use this information in future revision and expansion of the program.

Additional Information

SEE-North
University of Michigan Biological Station
Pellston, MI   49769
Contact: Dr. Mary Whitmore
(616) 539-8406

"SEE-North" does environmental education training in the Northern Lower and Eastern Upper Peninsulas of Michigan. Staff are currently working with the Northern Michigan Recyclers Cooperative of Gaylord, MI (Contact: Doug Hyde 1-800-342-6672) to organize a teachers' conference on recycling in the fall of 1988.

Michigan United Conservation Club
2101 Wood St.
Lansing, MI   48912
Contact: Mr. Kevin Frailley
(517) 371-1041

The Michigan United Conservation Club is a member of the Home Chemical Awareness Coalition (HCAC), an interagency group formed in 1985 to promote the proper use and disposal of common household products. HCAC has educational materials for fourth, fifth, and sixth graders entitled "Wastey Needs You". These materials are a series of fact sheets about products which are usually *not* accepted at hazardous waste collection days, and includes training for communities doing household hazardous waste collection. For more information contact:

Cooperative Extension Service
Michigan State University
404 Ag Hall
East Lansing, MI   48824-1039
(517) 355-9578

Ecology Center of Ann Arbor
417 Detroit St.
Ann Arbor, MI   48104
Contacts:
Jeryl Davis, Recycling Coordinator
Mike Garfield
(313) 761-3186

The Ecology Center of Ann Arbor is a non-profit organization that provides educational programs for adults and children, runs recycling programs, and lobbies for environmental concerns. It offers curriculum materials, primarily on household hazardous waste topics. They have revised the Golden Empire Health Planning Center curriculum entitled, *Toxics in My Home? You Bet!* to make it more Michigan specific. The revision includes a listing of other available educational materials. The Center also offers video tapes for teachers about the curricula, providing background and explanation for implementation. This video tape may be rented or copied out of state. Additional materials include a recycling education curriculum for grades K-8, a recycling slideshow for school-age children, and various materials on alternative pest control products and management techniques.

REFERENCES

Cotner, J., Natural Resources and Public Policy Agent, Genesee County Extension Service, Flint, MI, personal communication, August 23, 1988.

Frailley, K., MUCC, Lansing, MI, personal communication, August 23, 1988.

Garfield, M., Ecology Center, Ann Arbor, MI, personal communication, August 24, 1988.

Whitmore, M., SEE-North, Pellston, MI, personal communication, August 23, 1988.

Minnesota
Waste Management Board
Waste Education Program
1350 Energy Lane
St. Paul, MN   55108
Contacts:
Suzanne Hanson, Waste Education Program Manager
(612) 649-5786
Sue Thomas, Clearinghouse Coordinator (612) 649-5482
Mary Kay Newland, Research Analyst (612) 649-5483

Curriculum/Materials: In progress (K-12)
Waste Education topics
Teachers Guide, student lessons
Available: Fall/Winter 1988
Validated: Y

Solid Waste Management Background

The 1980 Minnesota Waste Management Act called for "reduction in indiscriminate dependence on disposal of waste." The 1984 draft of the *Hazardous Waste Management Plan* identified public education as an important component of effective waste management. Each county presents a solid waste master plan, which must now provide for local education measures. Minnesota does not have statewide mandatory recycling, but some communities have passed local recycling bylaws. In the spring of 1989, the legislature rejected major recycling legislation, which has removed the funding source for the Waste Education Coalition.

Educational Infrastructure

On June 1, 1985, the Minnesota Department of Education passed the "Elementary Education Rule", which became effective during the 1986-87 school year. The rule identifies integrated environmental education as one of ten required subjects for students in grades K-6. The types and specific content of the environmental education materials are decided by each Board of Education, which has considerable autonomy in determining curriculum programs.

Project History

In June 1985, the chair of the Waste Management Board appointed a 13 member citizens task force to address the waste education issue—the Waste Education Roundtable. The Roundtable produced a *Final Report* August 1, 1986, recommending the creation of a coordinating body to address the identified waste education needs. (The Waste Education Roundtable *Final Report* provides a thorough discussion of the variety of waste education activities in Minnesota; it is available from the Waste Management Board.) As a result, the Minnesota legislature created the waste education program under the Waste Management Board. The Chair of the Board appointed 15 members to the special task force, the Waste Education Coalition, with a budget of $190,000 for 1987-88. The members are representatives from public agencies responsible for waste management or public education. These agencies include the Minnesota Waste Management Board, Minnesota Pollution Control Agency (MPCA), Minnesota Department of Agriculture, State Planning Agency, Environmental Quality Board, and the Minnesota Environmental Education Board. Educational institutions, other public agencies, interested citizens and industry are also represented.

Project Description

The waste education program was established to "develop, implement and coordinate state and regional resources in an integrated, long-term waste education program which encourages the reduction, reuse, resource recovery, and proper management of solid and hazardous waste. (Waste Education Coalition, Mission Statement, November 23, 1987.) The identified audiences of such education include school audiences, residential populations, local and state government officials, waste management handlers, industry and others.

The Waste Education Coalition has three components—the K-12 school curriculum project, adult education, and a computerized clearinghouse of information. Three separate committees presently address the different aspects of the project; the overall goal, however, is a permanent, well-integrated system that offers materials and information to educators, provides citizens with knowledge about community programs, answers or refers waste management questions, and utilizes waste education activities throughout the state. The following description focuses primarily upon the curriculum aspects of the Minnesota program, but details about the other plan components are available from the Minnesota Waste Management Board. The failure of the recycling legislation has forced significant staff cuts, although the program has sufficient funds to work for another year on the curriculum. The Clearinghouse efforts will necessarily focus more upon local and state needs.

Curriculum Description

The first curriculum is for grades K-6, interdisciplinary, designed in a simple, single kit for thorough integration with existing curricula. The materials are designed to fit into the social studies and science framework of the state, and eventually may be incorporated into the Environmental Education Division materials. Actual materials include a Teacher's Guide and student lessons. The curriculum addresses topics of solid and hazardous waste management and attempting to balance waste subjects (e.g., recycling, landfilling, incineration). The curriculum is progressive through grade 6, with a future goal of developing a K-12 program. The concept of a "living example" is central to the curriculum—schools and waste education providers should exhibit sound waste management behavior and encourage students to become involved in local projects. For example, all curriculum materials are printed on recycled paper using nonhazardous materials. Teaching materials will be applicable to classroom, community and field settings. The Coalition recognizes that school audiences include students, teachers and administrators, and that each component of this audience has separate curriculum needs. It has therefore outlined the development priorities to adequately address those needs.

Curriculum Development and Review

The curriculum program began in the summer of 1987. The Coalition conducted a survey of Minnesota teachers (second and fifth grades) and identified that time limitations present the greatest obstacle to introducing new materials into the classroom. The goal, therefore, is to create flexible activities that are inherently integrated into the other required curriculum subjects, as defined under the Elementary Education Rule. The units must, however, be "hard core" enough to ensure a progressive curriculum structure.

The Clearinghouse Coordinator conducted a national search for curricula and other materials concerning solid and hazardous wastes, then evaluated and referenced them for selective use in the Minnesota curriculum. She estimates that the completed curricula will be 20-30% Minnesota specific. The legislative Roundtable also conducted detailed research of a few state (WA, NJ, VA, CA, D.C.) and foreign (Ontario) formal education programs, evaluating them by the ten guidelines previously identified as important to the Minnesota program. The Curriculum subcommittee determined the specific structure, goals and content of the curriculum.

The Coalition hired an educational consultant to oversee the actual curriculum development. This person is a curriculum development specialist and preferably has some waste background. The committee will encourage the consultant to utilize teachers, volunteers, waste specialists (technical and government), and educational officials in the development process. To review the curriculum, the Coalition hopes to involve environmental educators and volunteers on a committee once the curriculum program has become tangible enough to do so. The present date for completion is set for the fall of 1989.

Distribution and Teacher Training

The Waste Education Coalition structure is an internal state network of communication and distribution. Since many high-level individuals are involved in the curriculum process, each involved Department is aware of the program status and can be utilized to communicate with the public. The curriculum will be made available to educators upon a voluntary basis, free of charge after workshop sessions. Waste Management Board staff are considering hiring and training people from outside the agency to conduct workshops. The curriculum may also be stored and distributed by librarians and media specialists, organized by the central coordinating body, the Waste Education Coalition.

Testing and Evaluation

The Coalition plans to sponsor pilot programs in several communities when the curriculum is completed. Many teachers are aware of the project and are waiting to receive training and materials. Teachers will receive questionnaires after workshop sessions, to complete and return after implementation of the curriculum. The Waste Management Board staff also plans to conduct interviews with teachers to better evaluate the curriculum impact upon students, the usability of activities, and to what extent the curriculum is being integrated. The curriculum itself will include pre- and post-tests, "learner outcomes", to determine its impact upon the students. Optimally, a permanent curriculum overseer will supervise evaluations and arrange for any revisions. Results from evaluations will be processed and made available to individuals in a manner similar to previous studies sponsored by the Roundtable. The educational consultant may make additional suggestions for evaluation measures.

Current Status

The development schedule approximates completion date by fall/winter 1989.

REFERENCES

Countryman, L., Waste Education Program, St. Paul, MN, personal communication, June 20, 1989.

Hanson, S., personal communication, August 5, 1988.

Newland, M. K., Research Analyst, Waste Management Board, St. Paul, MN, letter, July 27, 1988.

Thomas, S., University of Massachusetts, Amherst, MA, personal communication, August 2, 1988.

Waste Education Roundtable: *Final Report,* Minnesota Waste Management Board, Crystal, MN, August 1, 1986.

Waste Education Roundtable: *Final Report: Appendices,* Minnesota Waste Management Board, Crystal, MN, August 1, 1986.

Missouri
Department of Natural Resources
Public Information Office
P.O. Box 176
Jefferson City, MO   65102
Contact: Stephen Schneider
(314) 751-3443

Curriculum/Materials: *Hazardous Waste Curriculum Guide* (7-9)
Groundwater Protection Curriculum Guide (4-12)
Some solid waste and related topics
Teachers Guide, student activities
Available: N
Free to Missouri teachers
Environmental Education Guides available

Solid Waste Background

Missouri has passed recent legislation concerning landfill regulation. There is no statewide recycling mandate, although some communities have ambitious voluntary programs. The Department of Natural Resources is conducting recycling feasibility studies.

Education Infrastructure

In 1985, Missouri passed the Excellence in Education Act, mandating specific objectives and testing for students in grades 2-12. The Department of Natural Resources works with curriculum guidelines to produce materials that teachers can effectively use in the classroom.

Project History and Description

The Department of Natural Resources works on materials independently of the Department of Education. They have received an EPA grant to develop the groundwater materials. They target only Missouri teachers, however, so materials are generally not available out of state.

Curriculum Development and Review

An oversight committee hired (stipend and expenses) seven Earth Science teachers to develop educational activities for the groundwater protection program.

Curriculum Description

The *Hazardous Waste Curriculum* guide outlines a five-concept program: defining waste, hazardous waste, exploring sources and effects, understanding responsibility for stewardship of the earth. The Guide provides teacher background information, specific objectives for each lesson, and suggested in-class activities.

Distribution and Teacher Training

The Department of Natural Resources utilizes a permanent distribution network to reach Missouri teachers with educational materials. The staff focuses the limited resources on teacher training workshops, where materials are provided for all participants. Teachers may receive college credit for attending workshops.

Testing and Evaluation

No testing and evaluation system exists specifically for the hazardous waste and groundwater materials. The Department does conduct general teachers surveys to evaluate what other services are needed; teachers are also surveyed concerning training workshops.

Current Status

Groundwater educational materials are available to Missouri teachers as of the spring of 1989.

Additional Information

The Missouri Department of Natural Resources adapted two out-of-state programs which were made available to Missouri teachers. Wisconsin teachers developed the Instruction-Curriculum-Environment (ICE) program, which consists of more than a dozen guides to different aspects of environmental education. A specific Environmental Education guide for each grade level consists of teacher guidelines and objectives, student activities, evaluation suggestions, and references to additional resources. The guides also provide supplemental curriculum material for existing subjects in the schools: *Life Science, Physical Science, Earth Science, Social Studies, Industrial Arts, Home Economics, Mathematics, Music, Art, Business Education,* and others. Iowa teachers developed the Energy Conservation Activity Packets for grade levels K-6. These guides are also supplementary and interdisciplinary, providing teacher objectives, suggested activities, worksheets, student evaluations, and helpful resource aids. A Review Committee, consisting of school teachers and administrators, and Department of Natural Resource staff, revised these two programs for Missouri use. These resources are no longer available through the DNR but can be obtained through the respective state Departments of Natural Resources/Environmental Protection.

REFERENCES

Brunner, F. A., Director, Department of Natural Resources, letter, August 17, 1988.

Schneider, S., Public Information Office, DNR, personal communication, August 29, 1988.

Schneider, S., letter, May 30, 1989.

New Jersey
Department of Environmental Protection
Division of Solid Waste Management
CN 414—401 East State Street
Trenton, NJ   08625
Contacts:
Mary Sue Topper, Project Specialist
(609) 292-0115
Vicki Kerekes, Program Development Specialist
(609) 292-0331

Curriculum/Materials: *Here Today—Here Tomorrow! Revisited* (4-8)
Solid Waste, Recycling topics
Teacher's Guide, student activities
Available: Y
Fee: Free, may change depending upon number of requests
Policy: Request permission to copy

Solid Waste Background
In April 1987, New Jersey passed a mandatory source separation and recycling act, requiring each county to submit recycling amendments to the original solid waste management plan by October 1987. These plans were to be approved by the Office of Recycling. Mandatory leaf composting legislation took effect in the fall of 1988. The Act also allocated funding for educational programs.

Educational Infrastructure
School districts set curriculum requirements in accordance with Department of Education guidelines.

Project History and Description
The original curriculum and its revision resulted from the cooperative efforts of the Departments of Energy and Environmental Protection, and the Conservation and Environmental Studies Center, with advisory support from the Department of Education. The revision currently underway is part of a new departmental initiative to provide a total education program on solid waste.

Curriculum Description
The original curriculum and its first revision were supplemental and multidisciplinary, designed to work flexibly with local school district curriculum requirements. The first two editions focused on energy, recycling and solid waste topics.

The current revision is extensive and will broaden the curriculum to cover all aspects of solid waste, such as incineration, recycling, landfilling, composting, and others. The Department of Environmental Protection describes the overall goals as follows: to help students acquire an awareness that a problem exists in the management of solid waste in New Jersey; to help students realize that they share a part of the solid waste problem; to help students acquire a basic understanding of the four-prong approach to solid waste management (source reduction, recycling, resource recovery, and landfilling) and to gain a working knowledge of the function and consequences of each; to encourage students to identify and implement specific actions consistent with the four-prong approach to solve the local solid waste problem. Specific objectives under each of the specific waste management topics are also included. The activities are structured with individual objectives. They include suggested subject areas, skill identification, materials required, a detailed procedural description, and extension possibilities where applicable. The Teacher's Guide also includes "copy me" pages, a glossary, a resources and references list including materials available from the Department of Environmental Protection, a solid waste fact sheet (for the U.S. and New Jersey), basic information about recycling, a "how to" section on school recycling programs, "About the Family" on New Jersey Solid Waste Mascots, and an evaluation form.

Curriculum Development and Review

The revised third edition was developed by New Jersey teachers and solid waste management professionals, including staff from the Departments of Environmental Protection, Division of Solid Waste Management, and the Office of Communications and Public Education. Teachers received a $50 stipend per day plus travel expenses to work on the project. The Department of Environmental Protection will make only the third edition available.

Distribution and Teacher Training

With the first edition, teacher training programs were presented in 11 counties, under the directorship of Robert D. Elder, Director of Curriculum for the Medford Public Schools. For the third edition, the Division of Solid Waste Management has two staff members who give curriculum workshops, and train teachers as facilitators to give workshops in their respective schools. The state will work with the County and Municipal Recycling Coordinators to encourage communication with local school districts. Coordinators are also trained to give workshops and provided with curriculum materials to distribute. Distribution of the new curriculum began in a few specifically targeted areas.

Testing and Evaluation

The third edition of the curriculum does not utilize extensive evaluation methods, but does include an evaluation form.

Current Status

Here Today—Here Tomorrow Revisited is available as of March 1989.

Additional Information

The Pennsylvania Department of Environmental Resources prints and distributes the original edition of *Here Today, Here Tomorrow* for grades K-12. Contact:

Pennsylvania Department of Environmental Resources
P.O. Box 2063
Harrisburg, PA   17120

REFERENCES

Here Today, Here Tomorrow, Conservation and Environmental Studies Center, Brown Mills, NJ, 1982.

Kerekes, V., Program Development Specialist, Division of Solid Waste Management, NJ DEP, personal communication, June 1988, August 3, 1988.

Kerekes, V., letter, March 10, 1989.

Topper, M. S., Project Specialist, Division of Solid Waste Management, DEP, letter, August 17, 1988. *(About the Guide.)*

New York
State Department of Environmental Conservation
50 Wolf Rd., Room 504
Albany, NY   12233
Contact: Frank Knight
(518) 457-3720

Curriculum/Materials: development stage
Module units on energy, natural resources, waste
(6-8)
Recycling Study Guide (4-8)
Recycling and Solid Waste topics
Ten activities
Available: Y

Solid Waste Management

In 1988, New York passed a new solid waste management act mandating that all local governments will implement recycling of marketable materials by 1992. Educational measures were stipulated as part of this legislation. Many towns already have local recycling resolutions, some of which have required source separation for many years.

Education Infrastructure

The State Education Department mandates a *State Syllabus,* defining specific curriculum requirements in each instructional area. The New York waste education program will provide a simple, easily integrated study guide to encourage teachers to infuse recycling and waste topics into the standard curriculum materials.

Project History and Description

The Science, Technology and Society Education Program (STS) has received more emphasis during 1988-1989 and has produced the first available curriculum materials on energy/natural resource topics.

The Department of Environmental Conservation began its educational efforts with work on an adaptation of the Wisconsin *Recycling Study Guide.* The funding for the program was part of the general allocation from the 1988 Solid Waste Act. The Department of Environmental Conservation's Environmental Education in the Division of Public Affairs assisted the Waste Reduction and Recycling Bureau which is responsible for the current waste education program. The Department utilized the State Education Department as a professional educational resource in the development of the *Recycling Study Guide* intended for teachers to use with their students. Completion date of this project is not determined.

Curriculum Description

The activities are module units designed for easy integration into middle school classroom subjects.

The *Guide* is being modeled after the Wisconsin *Recycling Study Guide,* produced by the Wisconsin Department of Natural Resources in January 1988. The New York materials will probably include ten activities, with background information and fact sheets.

Curriculum Development and Review

The STS committee consisted of representatives from the New York Department of Education, New York Energy Pool (utilities), State Universities of New York, and the Department of Environmental Conservation, which oversees the development of activity units for middle school students. The first unit on energy topics was completed during 1988-1989. During the summer, 40 teachers from around the state were brought in to evaluate and revise the proposed activities. The committee plans to produce four units per year; solid waste management will be the topic of the 1989-1990 program.

Work on the *Recycling Study Guide* began with a review of educational materials from other states, and selection of the Wisconsin guide as most adaptable to New York educational and waste management needs. An outline of the final *Recycling Study Guide* structure has been prepared, and two public school teachers hired to write classroom activities. All materials will be reviewed by the Solid Waste staff. Mr. Knight will work with the Environmental Education liaison at the Education Department to ensure that the materials will be easily integrated under New York curriculum requirements.

Distribution and Teacher Training

The STS program will include teacher training workshops to distribute and implement the educational materials. When the *Recycling Guide* is completed, teachers will receive a copy free upon request from the Department of Environmental Conservation. The Department will also utilize the Board of Cooperative Education Services (BOCES) system in the state, which has representatives in almost every county. This organization provides extra services to schools and offers 1-day teacher training workshops. The Environmental Education workshop can be adapted to include information about the *Recycling Guide.* The network will serve to distribute materials throughout the state.

Testing and Evaluation

The Department of Environmental Conservation will include an evaluation procedure, which has not yet been developed. The State Education Department will provide assistance with this component of the program. Evaluation information will be used to update and potentially expand materials in the future.

Current Status

A structure outline and draft of activities has been developed. Further development of the *Recycling Guide* was temporarily suspended; completion date is uncertain. The STS program will be introduced in the fall of 1989.

Additional Information

"Ithaca Recycles"
108 E. Green St.
Ithaca, NY   14850
(607) 273-3470

"Ithaca Recycles" is a recycling and educational resource for the Ithaca community, where mandatory source separation and recycling became effective in June 1988. Lynn Leopold, the Educational Coordinator, works directly with elementary and secondary schools teaching recycling. Her activities include presentations, slide and puppet shows, and paper making. She communicates regularly with the Department of Environmental Conservation.

Cornell University
Cooperative Extension Programs
Waste Management Institute
G20 Martha Van Rensselaer Hall
Ithaca, NY   14853-4401
Contact: Kenneth H. Cobb
(607) 255-1187

New York State Department of Environmental Conservation and the Cornell Waste Management Institute Cooperative Extension are developing a *Recycling Handbook*. This book will provide a compilation of recycling information to the general public, explaining recycling procedures, marketing, and listing state resources and other educational materials. The Institute also offers a listing of many solid waste educational materials available as of April 1988.

REFERENCES

Cobb, K., Cornell University, Waste Management Institute, Ithaca, NY, letter, August 25, 1988.

Cobb, K. H., *Solid Waste Education Materials,* April 1988.

Knight, F., NYSDEC, Waste Reduction and Recycling Bureau, personal communication, August 25, 1988.

Knight, F., personal communication, June 2, 1989.

Leopold, L., Ithaca Recycles, Ithaca, NY, personal communication, August 25, 1988.

Lewis, P., New York State Department of Environmental Conservation, Albany, personal communication, August 25, 1988.

Oregon
Department of Environmental Quality
811 SW 6th Ave.
Portland, OR   97204-1390
Contact: Alene Cordas
(503) 229-6046

Curriculum/Materials: *Rethinking Recycling* (K-12)
Waste reduction, recycling topics
Teaching aid, lessons
Available: Y
Fee: $15 (checks payable to DEQ)
Policy: Y

Solid Waste Background

Oregon enacted its Bottle Bill in 1971, the first such legislation in the U.S. In 1983, the Recycling Opportunity Act mandated that local governments provide recycling services for citizens. Citizens are not required to participate in recycling activities under this Act. The Department of Environmental Quality (DEQ) is responsible for planning and advocating solid waste strategies for the state. The DEQ is currently conducting studies to make further legislative recommendations.

Educational Infrastructure

The Department of Education requires that curricula follow general guidelines, but individual school districts can determine specific materials within those guidelines. Since the recycling curriculum is multidisciplinary and designed to cover up to a 2-week period, the Department of Education reviewed the materials but deemed an official validation unnecessary.

Project History and Description

The program began due to DEQ initiative, an allocation of approximately $20,000 from the Department budget.

Curriculum Description

The general goal of the program is to increase the recycling behavior of students, teachers and parents. The curriculum is a 1-week program, following social studies and science general sequences. Curriculum structure allows extension of the program into 2 weeks. The lessons are divided into grade level groupings, K-1, 2-3, 4-5, 6-8, 9-12. Each level includes activities and contains learning objectives, suggested teaching procedures, interdisciplinary activities, materials lists, optional lesson extensions, and masters for worksheets and overheads. The Resource Section contains background information, short facts and figures about waste reduction and related topics, several informational sheets and charts, a film list, glossary, and contact names in different communities.

The DEQ describes specific curriculum goals as: learning about the Bottle Bill and recycling opportunities in Oregon; using charts to compute energy savings around the world; considering how litter affects wildlife; considering the impact of buying habits on waste accumulation; role-playing landfill siting, and others. Many materials are designed as handouts which are taken home to families and communities. Activities can be done independently of the lessons. The materials also include "Guidelines for Recycling in the Schools", encouraging local recycling activities.

The curriculum is designed to be very simple and "teacher friendly". The DEQ hoped to make the curriculum attractive to all teachers, even those who are not particularly enthusiastic about recycling.

Curriculum Development and Review

The project began in 1985-86 with the formation of an Advisory Task Force consisting of environmentalists, educators, and representatives from the Departments of Education and Environmental Quality. This task force met over the course of $1^1/_2$ years to specifically identify Oregon's recycling education needs. The DEQ then advertised for professional curriculum development proposals. After reviewing the 20 respondants, the DEQ selected a consultant with strong educational background. The DEQ had compiled some materials from other states, resources from the Educational Resources Information Center (ERIC), and reviewed older Oregon materials.

The educational consultant developed most of the materials, which were then reviewed by the DEQ. All information was reviewed by the technical staff. Final editing was done by Ms. Cordas.

Distribution and Teacher Training

A volunteer organization, the Association of Oregon Recyclers, sponsored the six workshops and distributed curriculum materials in the summer of 1987, when the curriculum was first available. The DEQ did not have sufficient funds to offer other workshops. Workshops are offered as part of teacher in-service days. Project Learning Tree, an Oregon State University educational resource awareness organization, will feature and distribute the curriculum in eight scheduled in-services in various locations of the state during the 1988-89 school year.

The DEQ distributes the materials through the regional recycling coordinators in every county. Each coordinator is provided with a brochure explaining the materials, which is then communicated with the local schools. Oregon State teachers can obtain a free copy of the materials by sending an order form in to the DEQ.

The most effective distribution mechanism to date is a curriculum display at teacher conferences around the state. The Department of Education assists in identifying conferences, and Ms. Cordas attends these conferences to publicize the curriculum. Teachers in many different disciplines can see the materials and obtain order forms. The order form system allows the DEQ to keep records of requests. Ms. Cordas also gives presentations on the curriculum at statewide conferences.

Testing and Evaluation

The curriculum was field tested in 15 classrooms by volunteer teachers. These teachers were asked for suggestions and evaluation of the materials. The field testing resulted in additional activities provided by the DEQ.

An evaluation form is included in every curriculum packet, to be completed by teachers after curriculum implementation. These evaluations will be compiled and reviewed to provide a basis for curriculum revision.

Current Status

A second edition of the curriculum was prepared by Ms. Cordas and released in February 1988. It appears that a significant number of teachers are integrating at least a portion of the curriculum on a widespread basis.

REFERENCES

Cordas, A., Oregon DEQ, Portland, OR, personal communication, August 25, 1988.

Rethinking Recycling, Oregon Department of Environmental Quality, 1987.

Rhode Island
Department of Environmental Management
83 Park Street
Providence, RI 02903
(401) 277-3434
Contact: Cynthia LeVesque

Curriculum/Materials: *OSCAR's Options* (4-8)
Solid and Hazardous Waste Topics
2 volumes
Available: Y
Fee: $50 per volume, 2 volumes
Validated: N

Solid Waste Background

Twenty-nine of the thirty-two communities in Rhode Island use one central landfill, a facility owned by the state and operated by the quasi-state organization, Solid Waste Management Corporation. In 1986, Rhode Island passed a mandatory recycling law.

Educational Infrastructure

Rhode Island's 32 school districts have autonomous school committees; the Department of Education has no centralized authority to dictate specific curriculum requirements.

Project History

The bottle bill was repeatedly defeated in the Rhode Island legislature. The Ocean State Clean-up and Recycling Program (OSCAR) began as an effort to address Rhode Island's solid waste management crisis in lieu of a bottle bill, and was funded by a nickel tax on each case of carbonated beverages sold for distribution in Rhode Island.

In the spring of 1986, the Rhode Island legislature passed the Flow Control Bill, establishing the legal infrastructure for mandatory curb-side recycling. The bill also mandated the construction of three resource-recovery facilities and municipal recycling facilities (MRF). The Department of Environmental Management received authority to specify and enforce required recyclables provisions. Funding for *OSCAR's Options* comes from OSCAR budget; the total expenditure was approximately $100,000.

Project Description

The OSCAR program encompasses all aspects of Rhode Island's litter control and recycling effort. The Department of Environmental Management hired Ms. Carole Bell to develop educational materials that would aid in implementing the state program. The educational project began in the spring of 1985 with state-sponsored aluminum can recycling projects in approximately 12 grammar schools. The Department of Environmental Management provided modest money awards to school districts and educational materials to teachers. Ms. Bell conducted oral interviews with principals, teachers and children. The 6-week project seemed to be a strong success, based on perceived involvement and enthusiasm; however, once the competition ended, the apparent interest in recycling declined sharply. To encourage more consistent behavioral change and work toward longer-term solutions, the Department assigned Ms. Bell to develop a comprehensive waste management curriculum for grades 4 through 8.

Curriculum Description

OSCAR's Options is a two-volume, supplementary environmental education curriculum targeting grades 4 through 8. Volume I treats three topics: natural resources, litter, and household hazardous materials. The units in Volume II teach about the solid waste issues: landfilling, incineration, recycling, compost and source reduction. Each unit includes background information, detailed lesson plans in all subject areas (reading, language arts, science, social studies and math), vocabulary and bibliography, transparencies and supplementary brochures and magazines.

To encourage and facilitate integration of waste management lessons into existing school curricula, the OSCAR curriculum goal was to develop "easy-to-use" lesson plans that could function individually or as part of the larger comprehensive unit. The lessons would not require teachers to utilize resources other than those provided in the curriculum.

Curriculum Development and Review

Ms. Bell began the development process in the summer of 1985. The Department of Education provided her with a list of teachers known for their environmental interests. Information about the program was then sent to the network of environmental teachers in grade schools, indicating that the Department of Environmental Management would hire five people to work on the curriculum during the summer. Through an application and interviewing process, Ms. Bell selected five teachers, and assigned each one topics about which to develop lesson plans. These original topics were: landfills, recycling, litter, incineration, and household hazardous waste. The summer project proved partially successful, producing two curriculum units and resulting in the full-time employment of one teacher to develop the entire curriculum with Ms. Bell in the fall of 1985.

Ms. Bell conducted a national search for existing waste management curricula. The Departments of Education and Environmental Management (or equivalent) in each state received a request for curriculum materials or a reference to whomever could provide information. Ontario, Canada was also included in this search. Received materials were compiled and evaluated. Based upon information thus received, Ms. Bell and her co-author then developed Volume I of the curriculum over the following 6 months. Volume I includes three units: natural resources, litter, and household hazardous materials.

Units of the curriculum were sent to volunteer members/teachers of the Rhode Island Environmental Education Association for evaluation. The authors then incorporated comments into their revisions. The Department of Environmental Management printed 1,000 copies of Volume I in the spring of 1986.

Distribution and Teacher Training

Press releases described the OSCAR program and curriculum. Principals in all school districts and the environmentally oriented teachers received notices that the curriculum was available and workshops would be given in the schools upon request. Workshop participation was prerequisite to receiving a free copy of the curriculum. Ms. Bell responded to such requests and began to offer workshops and distribute curriculum copies.

Testing and Evaluation

After the first year of curriculum distribution, Ms. Bell sent questionnaires to all teachers who had attended workshops. The survey return rate was approximately 50%; Ms. Bell reports the response as ''positive''.

Ms. Bell developed, distributed and evaluated Volume II by the same procedures utilized for Volume I. The entire curriculum was completed by the spring of 1987.

Current Status

One person is allocated to the curriculum project; this ''maintenance'' person presents workshops for teachers and will potentially update the present materials.

Additional Information

Support from state and private agencies—The Department of Education provided the original teacher list for Ms. Bell, and was subsequently aware of but not actively involved with the project. The Department of Public Health was also aware of the project but not formally involved.

The Rhode Island Audubon Society supported the program by sponsoring a number of workshops and publishing information in their newsletter, as did other local environmental groups.

REFERENCES

Bell, C., University of Massachusetts, Amherst, personal communication, June 23, 1988 (conference presentation).

OSCAR's Options, description sheet, provided by the Rhode Island Department of Environmental Management, June 1988.

Bell, C., personal communication, Providence, RI, June 1989.

Tennessee Valley Authority
Knoxville, TN 37902
Contact: John M. Judy, Project Manager
(615) 632-6031

ENVIRONMENTAL/ENERGY EDUCATION PROGRAM

The Tennessee Valley Authority was created in 1933, comprised of the seven Tennessee River states. The TVA functioned in many capacities: constructing dams, forestry projects, and providing formal and informal educational opportunities. The regional nature of the agency created a unique opportunity for extensive cooperation with educational institutions in the area. TVA originally provided assistance to land-grant colleges and universities in the region, and in 1977 experimented with the first university-based center for the Environmental/Energy Education program. This first partnership with Murray State University in Murray, Kentucky proved highly successful, and soon expanded to include 11 other centers around the region.

Each center provides its "service area" with teacher training, regional services, technical assistance, program development and research. Together, the 12 centers function as a network and delivery system for environmental and energy education programs. Since 1977, approximately 25,000 teachers have received environmental/energy education training. More than 40 educational programs have been developed, many in the form of curriculum and instructional materials, such as *Waste: A Hidden Resource,* developed by Western Kentucky University. These materials are thoroughly distributed throughout the region.

The Centers provide a variety of regional services, ranging from technical assistance programs to information presentations. Center research focuses on educational materials; they perform field testing, impact evaluations, and surveys, for Center purposes and for outside contracts. Generally, Centers match TVA funds and progressively shift to greater dependence upon a variety of private and public financial sources. Eventually, the Center should become self-sufficient.

The nationwide Alliance for Environmental Education is utilizing the TVA university-based system as a model for a national network. The strategy is currently being implemented in the Mississippi River Valley (ten states). A document was produced providing guidelines for center establishment. (This document is available from TVA.) In summary, these recommendations suggest: a strong support base, needs assessment, center resource inventory, advisory committee, diverse funding sources, resource center, seminar/workshops programs, a strong community network, university environmental education programs, and student involvement. The TVA region includes 201 counties, the majority of which are reached through the university-based Environmental/Energy Education program.

The twelve university centers are:

University of Tennessee-Chattanooga
Memphis State University
Western Carolina University
Middle Tennessee State University
Western Kentucky University
University of Alabama-Huntsville
Tennessee Technological University
Tennessee Wesleyan College
Jackson State (Mississippi) University
North Georgia College
University of Tennessee-Knoxville
Murray State University (Kentucky)

REFERENCES

Judy, J. M., Project Manager, Environmental/Energy Education Program, TVA, Knoxville, personal communication, August 3, 1988.

Judy, J. M., letter, August 8, 1988.

Judy, J. M., and Hodges, L. M., "Fundamentals of Environmental Education," Knoxville, June 15, 1987.

Texas
Texas Education Agency
1701 North Congress Avenue
Austin, TX   78701-1494
Contact: Grace Grimes, Assistant Commissioner for Curriculum
and Instruction
(512) 463-9596

Curriculum/Materials: *Suggested Activities for Environmental Education in the Elementary Schools* (K-6)
Suggested Activities for Environmental Education in the Secondary Schools (7-12)
Solid Waste and Recycling Unit included
Science Framework (K-12)

Available: Y
Fee: $1 per copy
Policy: materials not copyrighted

Project History and Description

Texas teachers expressed a need for guidance in developing environmental education curricula. The Education Agency responded by developing these materials in 1977.

Curriculum Description

Each section includes background information, suggested student activities, and a topic outline. The materials were designed using a three-dimensional model consisting of the personal concerns of people, problems of environmental quality, and learning function of students.

Curriculum Development and Review

The Guides were developed by public school and college educators.

Distribution and Teacher Training

Teacher workshops are offered by Division staff at professional meetings and upon requests from schools.

Evaluation and Testing

Further information on evaluation and testing procedures is not available.

Additional Information:

Keep Texas Beautiful
P.O. Box 2551
Austin, TX   78768

Contact: Cliff McCormack
(512) 478-8813

Over 425 elementary schools in Texas presently use the *Keep Texas Beautiful* curriculum, a state-specific adaptation of the *Waste in Place* (K-6) curriculum developed by Keep America Beautiful, Inc. Keep Texas Beautiful serves as a clearinghouse for distributing the curriculum and related resources, and keeps statistics on curriculum use. In 1987-88, 189,296 elementary students were taught *Waste in Place*, through over 44,000 teacher hours. Keep Texas Beautiful offers grants to communities based on the success of the overall program including education, law enforcement, litter prevention, etc. Through this process they are able to obtain accurate evaluation data. The organization is independent of any state agency, but worked originally with the Department of Highways and Public Transportation on litter prevention, later expanding into the present program.

REFERENCES

Grimes, G., Texas Education Agency, August 19, 1988.

McCormack, C., Keep Texas Beautiful, personal communication, August 18, 1988.

Mendieta, H. H., Division of Solid Waste Management Director, Dept. of Health, letter, August 16, 1988.

Science Framework, Kindergarten-Grade 12, Texas Education Agency, 1987.

Suggested Activities for EE in the Elementary Schools, Texas Education Agency, 1977.

Suggested Activities for EE in the Secondary Schools, Texas Education Agency, 1977.

Texas
Department of Health
Division of Solid Waste Management
1100 West 49th Street
Austin, TX   78756-3199
Contact: Mr. Glendon Eppler
(512) 458-7271

Curriculum/Materials: *Project MORE: Meeting Our Responsibilities Effectively*
Solid Waste Management topics
Teachers Guide
Available: Yes, limited copies

Solid Waste Background

The Texas Department of Health, Bureau of Environmental Health, Division of Solid Waste Management is responsible for regulation of municipal solid waste management in Texas, primarily under the Solid Waste Disposal Act of 1969. In 1983, Texas passed the Comprehensive, Municipal Solid Waste Management, Resource Recovery and Conservation Act. Waste management strategies select resource recovery as preferable to landfilling. A 15-member Municipal Solid Waste Management and Resource Recovery Advisory Council is comprised of local government officials, interest groups and some citizens.

Educational Infrastructure

The Texas Education Agency produces *Essential Elements* (1984) which are mandatory requirements for every course taught at every grade level, from prekindergarten through 12th grade. Materials made available by the Agency are approved by the State Board of Education in regards to these requirements. Teachers instructing in particular courses must follow these guidelines. The Curriculum division of the Texas Education Agency develops materials in all subject areas; environmental education is generally coordinated by the Science staff. There is no Environmental Education requirement per se.

The science curriculum in Texas currently includes an Environmental Science course, which addresses the environment and waste management.

Project History and Description

The curriculum program began in 1981 as part of a public education campaign. The specific project was due to initiative within the Division of Solid Waste Management. Some funding was available through Subtitles C and D of the federal Resource Conservation and Recovery Act (RCRA); other funding came out of the Division budget.

Curriculum Description

Materials are supplementary and offered for voluntary integration into the classroom. The Teachers Guide is designed to be used with the multi-media presentation also called "Project MORE", which provides general information about solid waste management in Texas. The Guide contains a "Narrative" section, which gives extensive background information, "Student Activities", a glossary, and list of resources such as films and publications, and other agencies. The Department of Health has a film library which makes materials available at no cost to Texas teachers.

Curriculum Development and Review

The Teachers Guide was developed through an interagency agreement between the Texas Department of Health and the Edwards Aquifer Research and Data Center, Southwest Texas State University, under the directorship of Dr. Glenn Longley. Specific details of development are not available.

Distribution and Teacher Training

Materials were provided on an inquiry basis only. "Health Fairs" held in various locations around the state provided a communication mechanism. The Department of Health did not provide a statewide training program or attempt a statewide promotion campaign.

Testing and Evaluation

Not applicable.

Current Status

Only 12 copies available. Future printing and development of other materials depends upon funding sources.

Additional Information

Initially, the program coordinator planned to present the materials to the Texas Education Agency for approval. The Teachers Guide would then have been distributed through the Agency, with other educational materials. Funding was insufficient to complete this process.

REFERENCES

Eppler, G., Dept. of Health, Solid Waste Division, personal communication, August 22, 1988.

Teachers Guide: Project MORE, Division of Solid Waste Management, Dr. Edmund A. Marek and Dr. Robert D. Larsen, 1981.

Texas Dept of Health, Division of Solid Waste Management, "Programs Information: Texas Solid Waste Management," September 1987.

Vermont Department of Education
120 State Street
Montpelier, VT   05602
(802) 828-3135
Alan Kousen
Science Consultant

Association of Vermont Recyclers
c/o 55 East State Street
Montpelier, VT   05602
(802) 223-6009
Susan Pedicord

Curriculum/Materials: Teacher Resource Guide (K-12)
Solid and Hazardous Waste, Recycling
Student activities, Teachers Guide
Validated: Y
Available: Y, out-of-state policy not determined

Solid Waste Background

In 1987, Vermont passed solid waste management legislation calling for the following priorities in future waste management strategies: waste reduction, reuse, recycling, processing and land disposal. The Regional Solid Waste Management Districts in the state are responsible for developing and implementing waste management plans, in accordance with the Agency of Natural Resources regulations. The Solid Waste Bill created a landfill tax, funds from which support Solid Waste Division efforts and are available to Management Districts for a variety of programs. Recent legislation, Acts 78 and 200, have further empowered communities, agencies and private groups concerning solid waste management efforts.

Educational Infrastructure

The Department of Education acts in an advisory capacity to the local school districts. Curriculum requirements are autonomously set by those districts.

Project History and Description

The Association of Vermont Recyclers (AVR) initiated the curriculum program, approaching the Department of Education with a proposal in December 1987. Approval of this proposal began the consortium between the Department of Education, the Agency of Natural Resources, and AVR.

Curriculum Description

The curriculum materials are interdisciplinary, but most applicable to science, social studies and language arts. The Guide will contain at least 60 activities, divided into four grade level groups. "Copycat" pages for students will accompany the activities. The Guide will also contain background information about solid waste issues, emphasizing recycling and including household hazardous waste topics. A resource section will list other sources of information including publications, organizations, and solid waste management facilities.

Curriculum Development and Review

The curriculum development began with the formation of an advisory committee, comprised of three teachers, the science consultant from the Department of Education, a recycling expert from one Solid Waste Management District, and the Directors of Education from two Environmental Education organizations. Vermont teachers identified curriculum objectives and proposed activities. The committee served to oversee generally the development progress and insure that the program remained consistent with original goals and philosophy.

Ms. Pedicord, of the Association of Vermont Recyclers, began the activity development with a compilation of curriculum materials from around the country. She selected approximately 120 activities which seemed most relevant to Vermont needs. During the spring of 1988, AVR hosted a teachers workshop for the purpose of adapting and developing activities. Ms. Pedicord presented the activities she had selected, which were refined and altered. The teachers were divided into four groups, each responsible for one grade grouping; four solid waste experts were available to assist the groups. Teachers received reimbursement for expenses involved in workshop participation. Preliminary activities were field tested in approximately two dozen pilot classrooms. Activities were then revised, retested and submitted for a final revision.

Distribution and Teacher Training

The Association of Vermont Recyclers is responsible for teacher training workshops, which began in January 1989. Workshops are partially funded through the Department of Education program for science teacher training. The Department of Education will provide 700 guides for in-state use in each public and private school. A policy for out-of-state use has not yet been determined.

Testing and Evaluation

The curriculum materials were originally tested during the spring of 1988. Results from this test were incorporated into substantial revision during the fall teacher workshop. Pilot classrooms provided input for further revisions. The Association of Vermont Recyclers is responsible for testing procedures.

An evaluation form is included in the guide, to be returned to AVR by teachers after curriculum implementation. This form will also be distributed at teacher training workshops. AVR compiled and synthesized evaluation forms, which can be used in future revision of curriculum materials.

Additional Information

The Association of Vermont Recyclers has over 200 members, representatives from recycling groups and Solid Waste Management Districts, regional planners, and other individuals. It was incorporated in 1982 and primarily functions to produce a quarterly newsletter which updates solid waste management activities around the state.

Vermont Institute of Natural Science (VINS)
P.O. Box 86
Woodstock, VT   05091
Contacts:
Jenepher Lingelbach
Bonnie Ross
(803) 457-2779

Curriculum/Materials:	*Waste Away!* (5-6)
	Solid Waste, Recycling topics
	Curriculum Guide, student activities
Available:	Y
Fee:	18.95 plus shipping ($2.50)

The Vermont Institute of Natural Science (VINS) has developed an educational program on household solid and toxic waste, entitled *Waste Away!*. The purpose of the program is to provide information about current crucial waste management problems, and encourage changes in habits and lifestyles to help address those problems. The curriculum will be a four-session program, 2 hours per session, given at 2-week intervals involving teachers, volunteer parents, and VINS staff directly in classroom activities. The curriculum is organized into the following components: What is the waste problem? What are the roots of the problem? What are possible solutions? What are the "roadblocks" impeding such solutions? The curriculum ends with a "culminating festival", inspired by curriculum activities but not directly involving VINS staff. Ideally, this festival is a recycling day, planned and conducted by students, teachers, and parents, and involving the community. Curriculum materials include: the "mini-curriculum" on solid and toxic waste, a manual for classroom teachers and volunteers with information, activities and resources; materials for classroom use, fact sheets for families, and plans for the "Trash Festival" and follow-up activities. The Institute conducted a very successful pilot program in four Vermont schools during the spring of 1989. Institute staff will continue to work with four or five schools each semester; the curriculum materials are available for out-of-state use. In Vermont, *Waste Away!* complements the Teachers Resource Guide published by the Association of Vermont Recyclers.

REFERENCES

Lingelbach, J., Vermont Institute of Natural Science, personal communication, July 18, 1988.

Pedicord, S., AVR, Montpelier, VT, letter, August 1, 1988.

Pedicord, S., personal communication, August 29, 1988.

Ross, B., VINS, personal communication, August 17, 1988.

Ross, B., VINS, personal communication, June 2, 1989.

Ross, B., letter, June 5, 1989.

Waste A-Way!, information packet, VINS, July 1987.

Virginia
Office of Litter Prevention and Recycling
Division of Waste Management
James Monroe Building, 11th floor
Richmond, VA   23219
(804) 225-2667
Mrs. Lynn G. Hudson, Commissioner
Contact: Marcia R. Phillips, Acting Director of Education
(804) 786-5251

Curriculum: *Operation Waste Watch* (K-6)
Litter Prevention and Recycling
Curriculum, Teacher Guides
Available: For loan; can be reproduced with proper documen-
tation
Validated: Y

Solid Waste Background

Virginia does not have mandatory recycling, although many local communities
have started ambitious voluntary programs and a few have passed mandatory recycling by-laws. When the legislature began to consider deposit legislation in 1976,
the state decided to pursue an ongoing education program in lieu of a bottle bill.
One hundred and fifteen of the approximately 134 municipalities have organized
litter control programs, with a volunteer support staff organized by a local Litter
Control Coordinator. The education efforts are coordinated and supported by the
Division of Litter Control and Recycling.

Educational Infrastructure

The State Department of Education set forth the ''Standards of Learning Objectives'' in 1983-84, which require that certain subjects are addressed in K-6
curricula. Environmental Education is not included in these requirements. Local
Boards of Education have considerable autonomy to decide which curriculum materials are used to meet the standards. Ms. Phillips made a correlation between the
Standards of Learning Objectives and the *Operation Waste Watch* materials and
found that one quarter of science requirements were met by the curriculum.

Project History and Description

To support the educational program alternative to deposit legislation, the legislature created a Litter Control Tax. A gross percentage of sales is taxed from fast-
food, convenience and grocery stores, soft drink bottlers and beer distributors. One
half of the revenues are used to fund the Division of Litter Control and Recycling,
the remainder supports local Litter Control programs. The total annual budget is
approximately $1.2 million. The Division has certain guidelines under which the
local litter programs may apply funds to a variety of programs. The Division
encourages local litter control committees to fund year-round education programs.

Curriculum Description

The Virginia curriculum is sequential and multidisciplinary, offering seven levels (grades Kindergarten through six) which cover topics in recycling, litter control and solid waste management. The specific topics of each level are the following: Grade K: *Natural and Man-made Objects*—Students practice classifying natural and man-made objects indoors and out; Grade 1: *Waste Out of Place*—Students classify objects into "waste" and "useful", and learn to identify litter as well as dispose of their own trash properly; Grade 2: *Litter Pollution*—A study of the negative effects of littering: injury to humans and animals, environmental pollution, and waste resources. Students study a littered area, the characteristics of litter, and learn the right way to clean up. Grade 3: *Trash Trends*—Students trace historical changes in products and packaging, study the role of packaging in everyday life and learn the meaning of "biodegradable"; Grade 4: *Let's Waste Less Waste*—An examination of desirable changes to make in home waste disposal habits to reduce the amount of solid waste and litter. Also, an exploration of sources of litter in the community; Grade 5: *Trash Treasures*—Whereas many things are recycled by nature, most modern products must be recycled by people. Students learn how recycling and resources recovery can help save crucial resources often lost through carelessness; Grade 6: *Community Solutions for Solid Waste Pollution*—A study of one's locality and its solid waste problems along with the design of a model waste disposal plan for an imaginary community. The curriculum can be taught in a classroom or in community centers.

Activity cards are the core of the curriculum, both Individual Activity Cards (IAC) and Class Activity Cards (CAC). Each unit comes with a Teacher's Guide including a step-by-step instructional outline, a vocabulary list, and an index of resource materials. Units for grades K, 1 and 2 come with coloring books for children; units 3, 4, 5, 6 come with comic books. Each unit also offers filmstrips and audio cassettes, with a script in the accompanying Teacher's Guide to provide background. Each unit also has a colorful poster concerning a relevant theme to be used in the classroom or elsewhere. Each unit can last up to 5 weeks, but is extremely flexible for selective integration into whatever activities the teacher determines appropriate.

Curriculum Development and Review

The Division of Litter Control and Recycling developed the curriculum in 1979-80. The materials are based upon the *Waste in Place* K-6 curriculum and activities created by Keep America Beautiful, Inc. The Division also used the Washington State Environmental Education program as a model to guide curriculum development. Two individuals were hired to develop the materials. One professional originally worked with the Virginia Department of Education and worked in Environmental Science within the Education Department of the University of South Carolina. The other professional was from Old Dominion University, one of the state universities. They developed the sequential activities and submitted the package to the Division in 1979-80. The program was endorsed by the Department of Education, although not actively promoted or financially supported.

The curriculum was first promoted in 1980 by the Division of Litter Prevention working with the State Superintendent of Schools and the Virginia Education Association. Through many "hands-on" workshops given by Division staff, the curriculum was introduced into approximately 25 schools around the state as part of a pilot program. Many volunteer teachers evaluated and offered suggestions concerning the curriculum; the Division then incorporated these revisions into the packet. The curriculum was made available to all schools in 1981.

Distribution and Teacher Training

Operation Waste Watch is available at no cost to elementary and middle schools. The Division also utilizes school librarians and media specialists to advertise and distribute materials and information. Kits are usually kept in the library for teachers to use when convenient. Ms. Phillips exhibits the curriculum at State Science Teachers Conferences and State Education Association meetings. The Division staff originally offered workshops to both teachers and communities, but currently has developed other procedures. The local Litter Control Coordinators are trained by Division staff to teach curriculum workshops. The Division also created a video tape in 1986-87 that explains the curriculum and demonstrates teachers using the materials in a classroom. The tape is sent to schools upon request.

Testing and Evaluation

The curriculum itself contains pre- and post-tests to allow teachers to evaluate program effectiveness. The Division sends an evaluation to librarians to obtain a general sense of how widely the curriculum is used. Coordinators receive evaluations about curriculum quality and use; the return rate of such questionnaires is approximately 50%. The Division can also assess an increase in curriculum utilization by monitoring the number of coloring and comic books requested by schools. Each kit contains enough books for one class; students are expected to keep these materials. Presently, more than 80% of Virginia's elementary schools have *Operation Waste Watch* kits.

Current Status

Funding for the Division's education programs is ongoing under the Litter Control Tax. Currently, six professionals work within the agency to perform all of the litter control and recycling tasks. The curriculum has not been revised since the 1980 pilot program, but the Division hopes to revise the program to include more recycling activities. As a long-term goal, the Division would like to develop secondary-level curriculum materials working with the state universities. Currently, the Department promotes memberships in Ecology Clubs for junior and senior high schools students, and supports these autonomous organizations by offering awards for excellent projects.

The Division is also a member of the "Clean State Leaders Network", a national network of state peer groups formed in 1986 to facilitate the exchange of ideas and information relating to the control of litter. The 26 members meet annually to exchange information, and have recently expanded the original objective of litter prevention to include recycling.

REFERENCES

Operation Waste Watch, brochure, produced by Virginia Division of Litter Control and Recycling, Dept. of Conservation and Historical Resources.

Phillips, M., Acting Director of Education, Division of Litter Control and Recycling, Richmond, VA, letter, August 1, 1988.

Phillips, M., Virginia Division of Litter Control and Recycling, Dept. of Waste Management, Richmond, VA, personal communication, August 17, 1988.

Smith, H. L., Dept. of Education, Richmond, VA, letter, July 28, 1988.

Waste Education Roundtable: Final Report: Appendices (c), Minnesota, p. 49.

Washington
Department of Ecology
4350 150th Ave. NE
Redmond, WA   98052-5301
Contact: Jan Lingenfelter, Program Coordinator
(206) 867-7043

Curriculum/Materials: *A-Way with Waste* (K-12)
Solid and Hazardous Waste Management and Recycling
352 page curriculum
Available: Washington Department of Ecology
Expanded, revised edition available March 1989
Fee: $14 plus postage (UPS)
Policy: Materials may be duplicated with proper documentation*

Solid Waste Background

The state of Washington passed the Model Litter Control and Recycling Act in 1970-71. The revenue from taxing manufacturers, wholesalers, and retailers of products and packaging related to litter problems provides the Department of Ecology with approximately $6 million annually. The Department of Ecology has traditionally been responsible for controlling litter, promoting recycling, and educating citizens about the problems of waste management.

Several recent mandates pertaining to solid waste and education require that the Department provide public education. These responsibilities have been reallocated to a new office, the Office of Waste Reduction, Recycling and Litter, a nonregulatory body which will pursue waste management reduction techniques. The educational programs will include not only solid and hazardous wastes, recycling, and litter control, but also draw upon the water and air quality resources of the Department.

Washington does not have mandatory recycling or a bottle bill. Recent studies, however, have shown increased recycling in the state; recycling activities appear to be shifting to curbside pick-up programs.

Educational Infrastructure

There is no legislative mandate requiring that environmental education be included in school curricula. The Superintendent of Public Instruction Office does have an Office of Environmental Education which develops and promotes environmental educational materials. In 1985, an interagency Environmental Education Committee was created; the Department of Ecology and Office of Environmental

* This material borrowed from *A-Way With Waste* curriculum guide. A program of the Washington State Department of Ecology.

Education have representatives on this committee, whose function is to better pursue environmental education opportunities in the state. The Department of Ecology works closely with the Office of Environmental Education and the Department of Education.

The Department of Education does not mandate any specific requirements which must be followed in curriculum development, but does review and validate all Department of Ecology materials.

Project History and Description

The Department began work on the *A-Way with Waste* curriculum program in 1981. It is the core of the public information program. The relevant legislation and funding initiatives are described in the "Solid Waste Background" section.

Curriculum Description

The primary goal of the K-12 program is to promote a reduction in waste and an increase in recycling in school and at home. The guide includes 80 activities, current articles and research reports, a solid waste fact sheet, glossary, bibliography, list of resources, and an activity evaluation form to be returned to the Department of Ecology. The curriculum is organized around four key concepts: revise, reuse, recycle and recover, prioritized in that order. The activities are multidisciplinary and listed by concept, subject, and discipline. Each activity includes title, rationale, suggested grades, learning outcome, teacher background, necessary materials, learning procedure, pre- and post-test questions, bibliography and additional resources. The activities encourage students to revise awareness and attitudes about what they throw away, to change their waste disposal habits, to reuse and conserve resources, and to participate in recycling activities. The materials are designed to give students a sense that waste is a problem they can help solve. Options for school recycling programs are included in the guide.

The revised edition is even broader in scope, including topics in solid and hazardous waste, recycling, litter, Superfund sites, and others.

Curriculum Development and Review

The curriculum was developed in 1981 through a cooperative effort of the Association of Washington School Principals, the Washington Education Association, Superintendent of Public Instruction's Office, particularly the Washington State Office of Environmental Education, and the Department of Ecology. The Department of Ecology compiled materials from other states through an Educational Resources Information Center (ERIC) data base search. Elementary and secondary educators and principals from around the state contributed activities to the curriculum. Staff from the Superintendent of Public Instruction Office, solid and hazardous waste managers, recyclers, citizens groups and Department of Ecology staff gave presentations to help teachers develop activities.

The Department of Ecology edited, organized and expanded the materials using other materials and information. The personnel involved in development included a coordinator, writers/editors, a senior advisor, technical assistants, review committee members, clerical support, research assistants, illustrators, layout designers, workshop facilitators, and individuals from the Washington State Recycling Association.

A committee of recycling industry and environmental organization representatives, waste managers, and environmental educators reviewed the curriculum during its first year.

Distribution and Teacher Training

The curriculum is voluntary and available at no cost to Washington teachers after attending the required training workshop. Promotion and distribution began in 1984. The Department of Ecology trains staff to conduct workshops throughout the state on the *A-Way with Waste* program. Waste managers and recyclers teach workshop participants about local waste situations. Twenty teachers participate in each workshop; Washington has approximately 30,000 primary and secondary school teachers. The Department of Ecology pays substitute fees for all teachers who attend; workshops can also serve for college credit and in-service days. The Department has 5 workshop facilitators and 11 staff members.

A communication network exists between schools and the Departments of Education and Ecology. In addition, an "A-Way with Waste: News for Schools" newsletter is sent to all teachers who attend training workshops. Follow-up contact is greatly encouraged; teachers can receive additional free materials, borrow audio-visual resources, or find answers to any solid waste and education questions.

Testing and Evaluation

After the review process, the materials were field tested under a grant from Snohomish County. A series of ten teacher-training workshops involved approximately 150 Snohomish County teachers.

All teachers receive an evaluation form at the end of the training workshop. Ms. Lingenfelter also conducts a telephone survey every year of approximately 20% of the teachers who have attended workshops. She has a standard questionnaire which addresses teacher needs, response to curriculum suggestions, success of school recycling programs, etc. All information from evaluations and surveys is compiled to provide a basis for further curriculum revision. The curriculum has been revised twice since 1981.

Current Status

The *A-Way with Waste* education program is now under the jurisdiction of the Office of Waste Reduction, Litter Control and Recycling. The expanded, revised edition is available after March 1989.

REFERENCES

A-Way with Waste: A Waste Management Curriculum for Schools, 2nd ed., Washington State Dept. of Ecology, July 1985.

A-Way with Waste, Dept. of Ecology brochure, Washington State Department of Ecology.

Lingenfelter, J., letter, June 10, 1988; personal communication, August 24, 1988.

Wisconsin
Bureau of Information and Education
Department of Natural Resources
P.O. Box 7921
Madison, WI 53707
Contact: Anne Hallowell
(608) 267-5239
Joel Stone, Recycling Education Coordinator
Bureau of Solid Waste Management
(608) 266-2711

Curriculum/Materials: (DNR)
Recycling Study Guide (4-12) 32 pp.
Solid Waste and Recycling topics
Available: Y
Fee: one copy free
Donations appreciated for large requests
Policy: Materials not copyrighted
May be reproduced or borrowed, credit requested

Solid Waste Background

Wisconsin developed a statewide Solid Waste Recycling Authority in 1973 to build and operate regional resource recovery facilities. The Authority was not able to develop successfully these facilities and was disbanded in 1983. Simultaneously, local government was given additional powers to develop recycling/resource recovery facilities while the Department of Natural Resources was given the responsibility of providing information, education and technical assistance. The state has also adopted a hierarchy of preferred waste management techniques—reduction, reuse, recycling, composting, waste-to-energy, landfilling—and has adopted requirements for communities to develop collection facilities for motor oil, plastic, glass, aluminum and newspaper, and has recently enacted a prohibition of the landfilling of yard waste as of January 1, 1993.

Educational Infrastructure

In 1986, Wisconsin enacted legislation requiring that Environmental Education be infused into all curriculum subjects. This legislation expanded the previous conservation requirement. The Department of Public Instruction published a *Guide to Curriculum Planning in Environmental Education* to assist with this integration. (The *Guide* is available through the ERIC system or through the Wisconsin Department of Public Instruction.) The Department of Public Instruction does not develop actual curriculum materials. Legislation also outlines more rigorous pre-service teacher training in environmental education.

Project History and Description

The Department of Natural Resources *Recycling Study Guide* project began in

1985 as an in-house effort to address perceived needs for waste education. The Bureaus of Solid Waste Management (SWM) and Information and Education (I&E) worked cooperatively on the project. Information and Education funded development, review and distribution; Solid Waste Management funded printing. Team members involved from I&E are graduate-level trained professionals in science and environmental education. Solid Waste Management staff served as technical advisors.

Curriculum Description

The *Recycling Study Guide* is intended to help teachers and students understand what solid waste is, where it comes from, why it is a problem and what can be done about it. The *Guide* includes an overview of solid waste and recycling, a glossary, suggested activities and a list of resource publications, audio-visual materials, and organizations. It is designed to stand alone, yet complements the Wisconsin Department of Natural Resources Magazine, *Special Recycling Edition* (available from DNR).

Curriculum Development

After initial meetings to determine key concepts and a general project philosophy, each team member wrote an activity, which was submitted to Ms. Hallowell for editing. Some materials are drawn from the Washington State *A-Way with Waste* curriculum. Ms. Hallowell edited and restructured the activities to create a consistent, teacher-friendly guide. The activities were then sent back to each original creator for suggestions and review. A few individuals involved with recycling education outside of the Department of Natural Resources were asked to review the materials. The *Guide* was printed in early 1988. Suggestions and comments are welcome from any individuals using the materials.

Distribution and Teacher Training

A copy of the *Guide* was sent to all schools, nature centers, University of Wisconsin Cooperative Extension offices, community recycling programs, Land Conservation Department offices and many others. The Bureau communicates with all Wisconsin schools through an ''Environmental Education News'' newsletter, received by each principal; over 8,000 individual environmental educators also receive the newsletter. The Department of Public Instruction also publicizes Bureau Information. Original publicity for the *Guide* included press releases, radio announcements, and a mailing to all legislators.

The Bureau of Information and Education is offering recycling teacher workshops, beginning in the fall of 1988. The Bureau is also conducting Recycling Workshops around the state for recycling resource personnel (manufacturers, legislators, extension agents, community recyclers and so forth). The goals of these workshops are to facilitate cooperation among recyclers and provide them with materials and methods for educating teachers and the public. The Bureau has four professionals in its Central office, and two regional educators.

Testing and Evaluation

The *Recycling Study Guide* does not have any built-in evaluation measures, nor does the Bureau of Information and Education have enough staff to collect and analyze data. The developers are receiving general feedback and plan to incorporate these suggestions into future revisions.

Additional Information

The Bureau of Information and Education appreciates donations from any agency or organization that requests more than one copy of the *Recycling Study Guide*. These funds will support future printing and distribution. Donations should be made out to: Education Programs, Wisconsin Department of Natural Resources.

The Department of Natural Resources offers many public information and education publications on recycling and resource recovery, and other aspects of natural resources. Bibliographies and order forms of these resources are available from:

Education Section
Bureau of Information and Education
Department of Natural Resources
P.O. Box 7921
Madison, WI   53707

Wisconsin
Environmental Resources Unit
University of Wisconsin-Extension
Madison, WI   53707
Contact: Elaine Andrews
(608) 262-0020

Curriculum/Materials: (ERU)
Leader Training Program
Annotated Bibliography on Hazardous Materials Education
Outline of Subtopics of Household/Hazardous Waste
Hazardous and Toxic Materials topics
Available: Y
Fee: minimal cost (duplication and postage)
Policy: see specific materials

Project History and Description

The University of Wisconsin Environmental Resources Center originated in the late 1960s as part of the State Cooperative Extension Service. The five-campus

Center develops formal and non-formal educational materials, offers workshops to teachers and community leaders, teaches credit courses for educators and publicizes environmental education courses offered through other institutions. One course specifically addresses local environmental problems, focusing on hazardous materials; the credit fee covers both instruction and distribution of educational materials to teachers for classroom reference and use. Recent public instruction curriculum requirements include K-12 environmental education opportunities. This legislation encourages teachers to incorporate waste education into the applicable subjects. Ms. Andrews produced a document correlating the Department of Public Instruction curriculum guides for Health, Science and Social Studies to key ideas that could include environmental topics. The Department of Public Instruction reviews and authorizes the workshops that the Extension offers for teacher credit renewal.

Curriculum Description

The Extension Service materials include a Leader Training Program which combines workshops and guidelines concerning toxic and hazardous materials. These are generally offered for non-formal educators, but teachers may participate as well. The annotated bibliography list educational materials on hazardous waste. Materials listed include curricula and audio-visual resources. The outline of hazardous waste subtopics is designed to help teachers easily integrate waste education into the many different subjects they are required to teach.

Distribution and Teacher Training

The Environmental Resources Center publicizes materials through mailings to school superintendents, university mailing lists, newsletters, and through production of catalogs and brochures. The "A to Z of Environmental Education" brochure provided to teachers gives information on credit course offerings. Materials are often shared and distributed in these courses. The Center has approximately 20 faculty members who offer workshops on an occasional basis.

Testing and Evaluation

The Environmental Resource Center uses a self-report evaluation tool and interacts personally with many individuals that use materials, attend workshops or take courses. Perceived responses and needs for new materials are incorporated into Extension activities. Evaluation summaries for some programs are available.

REFERENCES

Andrews, E., Environmental Resources Unit, personal communication, August 19, 1988.

Grover, H., State Superintendent of Public Instruction, letter, August 10, 1988.

Hallowell, A., Bureau of Education and Information, personal communication, August 19, 1988.

Recycling Study Guide, WI Department of Natural Resources, January 1988.

EDUCATIONAL RESOURCES INFORMATION CENTER (ERIC)

ERIC is an information system sponsored by the Office of Educational Research and Improvement, within the U.S. Department of Education. It collects, abstracts, indexes, and distributes printed materials in all areas of education through many different services, which include: databases, abstract journals, microfiche, computer searches, on-line access, document reproductions, information analyses and syntheses, and others. ERIC consists of a central Government Office, 16 subject-specialized "clearinghouses", a central editorial and computer facility, a central "ERIC Document Reproduction Service", and a commercial publisher.

The ERIC system is computer-searchable, through keywords, authors, and other standard search elements. All Clearinghouse materials are referenced through the same indices. On-line retrieval services are available through three different vendors: DIALOG Information Services, System Development Corporation (SDC), and Bibliographic Retrieval Services (BRS) and CDROM. Approximately 95% of referenced materials are available on microfiche through the ERIC Document Reproduction Service (EDRS). Over 750 organizations subscribe to the EDRS and receive all microfiche materials. These organizations are referenced geographically in the *Directory of ERIC Information Service Providers,* which is available free from the ERIC Processing and Reference Facility.

ERIC references printed materials in all areas of education, including "fugitive" documents; curriculum materials often are such documents, not accessible through any standard index system. Referenced documents can be accessed through the database, and are also indexed monthly through *Resources in Education* (RIE). Journal articles are announced through *Current Index to Journals in Education.*

Each clearinghouse has the responsibility to develop and maintain contacts in its particular field, to aid in the primary task of identification and collection of materials. Such contacts are established with organizations that produce education materials, such as colleges and universities, federal governmental agencies, state agencies, specific school districts, professional organizations, etc. Each clearinghouse develops special publications, such as information analysis products, information bulletins, short "digests", monographs, bibliographies and others. Publication lists are available from the respective clearinghouse upon request.

The Clearinghouse for Science, Mathematics, and Environmental Education (SMEAC) is most relevant to waste management education and related topics. Since 1971, SMEAC has had specific responsibility to provide services concerning environmental education. ERIC/SMEAC has produced more than 90 special publications concerning environmental education, such as *One Hundred Teaching Activities in Environmental Education.* A publication list is available upon request.

For more information contact:

Dr. John F. Disinger
Associate Director
ERIC/SMEAC
The Ohio State University

1200 Chambers Rd., Room 310
Columbus, OH   43212
(614) 292-6717

ERIC Document Reproduction Service (EDRS)
3900 Wheeler Ave.
Alexandria, VA   22304
(703) 823-0500
(800) 227-3742

ERIC Processing and Reference Facility
ORI, INC. Information Systems
4833 Rugby Ave., Suite 301
Bethesda, MD   20814
(301) 656-9723

U.S. Department of Education
Office of Educational Research and Improvement (OERI)
Educational Resources Information Center (ERIC)
Washington, DC 20208
(202) 254-5500

Clearinghouses:

Adult, Career, and Vocational Education
Ohio State University
1960 Kennedy Rd.
Columbus, OH   43210
(614) 486-3655

Counselling and Personnel Services
University of Michigan
School of Education, Room 2108
610 East University St.
Ann Arbor, MI   48109
(313) 764-9492

Educational Management
University of Oregon
1787 Agate St.
Eugene, OR   97403
(503) 686-5043

Elementary and Early Childhood Education
University of Illinois
College of Education
805 W. Pennsylvania Ave.
Urbana, IL   61801
(217) 333-1386

Handicapped and Gifted Children
Council for Exceptional Children
1920 Association Dr.
Reston, VA   22091
(703) 620-3660

Higher Education
George Washington University
One Dupont Circle, N.W., Suite 630
Washington, D.C.,   20036
(202) 296-2597

Information Resources
Syracuse University
School of Education
Huntington Hall, Room 030
Syracuse, NY   13210

Junior Colleges

RESPONDENT STATES WITH NO WASTE EDUCATION PROGRAMS*

Alaska   Alaska Department of Education
                                            P.O. Box F
                                            Juneau, AK   99811

The state of Alaska has no waste education program, but the Municipality of Anchorage has hired a specialist to develop a hazardous waste curriculum. The program is primarily geared toward fifth and sixth grade students, but is appropriate for high school students and adults. The final product was scheduled for completion in the spring of 1989.

Reference
Vande Visse, E., Palmer, AK, letter, September 4, 1988.

Arkansas   Department of Pollution Control and
                                            Ecology
                                            8001 National Drive, P.O. Box 9583
                                            Little Rock, AR   72209

Status
The Department of Pollution Control and Ecology does not have a waste education program. The Environmental Academy at Southern Arkansas University, Technical Station, offers adult courses in solid waste management, and will soon offer special student group courses. Educational materials on water are available. (Environmental Academy, Southern Arkansas University-Tech Station, Camden, AR   71701.)

References
Anderson, D., Arkansas Environmental Institute, personal communication, August 25, 1988.
Etchieson, D., Senior Planner, Solid Waste Division, Little Rock, AR, letter, August 5, 1988.

*Some states may have individual educational materials.

Delaware Solid Waste Authority
P.O. Box 455
Dover, DE 19903-0455

Status

Delaware does offer a Coloring/Story book series on some waste issues. The books available are: *The Story of Trash Collection, Where Does Our Trash Go?*, and *Take a Trip With Me Through the Delaware Reclamation Project*. All coloring books feature "Trash Can Dan", a "copyrighted waste buff". These materials are available to other states, and can be adapted for other educational programs. The Governmental Refuse Collection and Disposal Association (GRCDA) has printed a large number of the solid waste coloring books. Delaware does not have any other waste education materials.

References

Take A Trip With Me Through the Delaware Reclamation Project, Delaware Solid Waste Authority, Dover, 1984.

The Story of Trash Collection: A Coloring Story Book, Delaware Solid Waste Authority, Dover, 1987.

Vasuki, N. C., General Manager, Delaware Solid Waste Authority, Dover, letter, July 27, 1988.

Where Does Our Trash Go? The Story of the Central and Southern Solid Waste Facilities, Delaware Solid Waste Authority, Dover, 1985.

District of Columbia Office of Instruction
Instructional Services Center
Langdon Elementary School
20th and Franklin Streets, N.E.
Washington, D.C. 20018

The District of Columbia Public Schools have a variety of curriculum guides and educational materials. Recent budget constraints have made revision and re-printing impossible. The materials are therefore not available to any individuals or organizations outside of the District.

Reference

Harbeck, M. B., Acting Director, Instructional Services Center, Washington, D.C., letter, September 1, 1988.

Kansas   Department of Education
                            120 East 10th St.
                            Topeka, KS   66612S

Status

Kansas does not have any specific waste education materials, but does have a comprehensive Environmental Education program. The Kansas Advisory Council for Environmental Education (KACEE) was created in 1969, comprised currently of more than 78 private and public organizations, institutions, and businesses. In 1978, the State Board of Education officially recognized KACEE, and authorized the Department of Education to work in cooperation with the Committee to develop and implement a comprehensive state plan for Environmental Education. Some waste education topics are infused into other areas of Environmental Education.

References

Anshutz, R., Program Specialist for Science, Department of Education, Topeka, KS, personal communication, August 29, 1988.

Leighty, R., Acting Director, Educational Assistance Section, Department of Education, Topeka, KS, letter, July 29, 1988.

Maryland   Department of Education
                            200 West Baltimore St.
                            Baltimore, MD   21202-2595

Status

The Maryland State Department of Education has not developed materials related to solid waste management issues. With Department support, local school systems have developed curricula and conducted teacher in-service programs that address watershed issues, including waste management topics.

Reference

Heath, G., Environmental Education Specialist, Maryland State Department of Education, Baltimore, letter, August 25, 1988.

Mississippi   Department of Education
                            P.O. Box 771
                            Jackson, MS   39205-0771

Status

Mississippi has not developed a solid waste curriculum.

Reference

Brewer, R., Director, Bureau of School Improvement, Department of Education, letter, August 12, 1988.

New Hampshire   Department of Environmental Services
Waste Management Division
6 Hazen Dr., P.O. Box 95
Concord, NH   03301

New Hampshire does not have mandatory recycling statewide, but many towns have recycling bylaws or voluntary programs. The state is currently updating its solid waste management program. Recent legislation places strong emphasis on training personnel involved with solid waste management.

The Department of Education includes waste management topics as part of secondary-level curriculum materials concerning agriculture. Relevant topics include composting, soil and water management, land use, hazardous waste care and disposal, soil fertilization and the general environment. The Department hopes to develop some waste education materials this year.

References
Howard, A. H., Commissioner, Dept. of Environmental Services, Office of the Commissioner, letter, July 26, 1988.

Minichiello, J., Director, Waste Management Division, personal communication, August 4, 1988.

Mitchell, M. L., Curriculum Supervisor, Agricultural Education, Dept. of Education, Concord, NH, letter, August 24, 1988.

Taylor, A., Environmental Research Group, personal communication, August 5, 1988.

New Mexico   Department of Education
Education Building
Santa Fe, NM   87501-2786

Status
New Mexico does not use any waste education materials. The state does use some other environmental education materials such as *Project Learning Tree, Project WILD,* and some materials produced by agencies such as the National Wildlife Federation and the United States Forest Service.

Reference
Graham, B. K., Science and Conservation Consultant, Dept. of Education, letter, August 4, 1988.

North Carolina   Department of Public Instruction
                                    Education Building
                                    116 West Edenton St.
                                    Raleigh, NC   27603-1712

Status

The science curriculum in North Carolina includes several environmental concepts, but does not concentrate on specific topics such as waste education.

Reference

Taylor, P. H., Director, Division of Science, DPI, letter, July 28, 1988.

Pennsylvania   Department of Environmental Resources
                                    P.O. Box 2063
                                    Harrisburg, PA   17120

Pennsylvania used the New Jersey "Here Today, Here Tomorrow" curriculum, and offers a small packet of Recycling Lesson Plans. The Department of Education did not provide information concerning any comprehensive state environmental education program.

Reference

D'Elia, E., Solid Waste Program Specialist, Department of Environmental Resources, Harrisburg, PA, letter, September 12, 1988.

South Carolina   Department of Education
                                    Rutledge Office Building
                                    Columbia, SC   29201

                                    Dept. of Health and Environmental
                                      Control
                                    2600 Bull St.
                                    Columbia, SC   29201

Status

The Department of Education has not sponsored or developed any curriculum materials concerning solid waste topics. The Department of Health and Environmental Control does not have any curriculum materials on environmental education.

References

Talton, E. L., Science Consultant, Basic Skills Section, Department of Education, letter, July 28, 1988.
Truesdale, H., Chief, Solid and Hazardous Waste, Department of Health and Environmental Control, letter, August 3, 1988.

South Dakota Department of Education and Cultural Affairs
700 Governors Dr.
Pierre, SD   57501-2293

South Dakota does not have an environmental education specialist within the Department of Education. The Department of Natural Resources did not offer any information concerning recycling curricula.

Reference
Hansen, J. O., Secretary, Department of Education and Cultural Affairs, Pierre, SD, letter, September 13, 1988.

Results

The categories that distinguish between different state programs are synthesized in Table 1. These categories provide an analytical structure for comparative evaluation of state approaches to solid waste management, specifically concerning recycling efforts and educational aspects of management strategies.

Materials Available

All of the states included in the table offer educational materials to teachers within the state. Three quarters offer materials out-of-state as well, usually for no cost or for a minimal fee. All of the educational materials referenced can be reproduced for classroom use. Copyright information is provided for guidance on the use of materials such as the adaptation of materials and incorporation into a new curriculum. Requests for materials from other states should be limited to programs previously identified as relevant and therefore likely to be implemented in a curriculum program.

Solid Waste Management Background

Since waste management strategies differ between states, solid waste management information allows outside agencies to identify states with similar technical strategies and evaluate its respective approach to solid waste/recycling education. The materials selected, developed and distributed by the state will inevitably reflect its environmental and educational needs; such data are therefore critical to an accurate assessment of the curriculum program. During the past 5 years, solid waste legislation has proliferated, particularly statutes requiring some form of mandatory recycling in the 1990s. In a few cases, such as Florida, mandatory development of educational materials is part of solid waste legislation. Most legislation identifies education as a desirable component of an integrated waste management plan. A number of states—Florida, Indiana, Michigan, Minnesota, Rhode Island, Virginia, Washington—allocated specific funds from legislation or tax programs to educational development. Some of these states—California, Minnesota, Oregon, Virginia and Washington—began their legislative programs considerably earlier than the rest of the country. The extended development period is reflected in the comprehensive nature of their educational programs.

Educational Infrastructure

Curriculum structure and content are developed within the educational parameters of particular states. Curriculum requirements or standardized testing measures must be considered in an overall evaluation of the education program, especially with respect to the transfer of materials. Research identified the broad educational infrastructure, pinpointing specifically the relationship between Departments of Education and local school boards.

Although Department function ranges from advisory to regulatory, most local school districts enjoy relative autonomy to determine specific curriculum requirements, within broad guidelines set by the state. Seven of the respondent states mandate more restrictive curriculum parameters, two of which require integration of environmental education (EE) materials into classroom activities (WI, MN). The EE mandate does not specify solid waste or recycling topics for integration, and no clear correlation exists between EE requirements and comprehensive waste management education programs. Classroom integration of specific recycling education materials is universally voluntary.

Project History and Description

Project background information is divided into several subcategories to provide a broader context for program evaluation. Different sources of initiative in each state are correlated with the creation of varied institutional structures to pursue curriculum development. All of the respondent states referenced in Table 1 have some form of state-sponsored education program, although sponsorship ranges from $20,000 Department of Environmental Protection allocations to educational budgets in excess of six million dollars.

In the majority of programs, one or two individuals within the Solid Waste Management Division are responsible to collect, synthesize, develop and distribute educational materials in addition to their other projects. Isolated resource allocation lengthens the development process and inhibits the scope of materials search, distribution, teacher training, pilot and field testing, evaluation, review and revision. The conduct and success of such programs are precariously dependent upon particular individuals and continued sources of funding from fluctuating state budgets. Although many of the isolated programs have the potential to produce effective activities, the materials comprise only one component of an implementation program that is critical to influencing the goals of increased awareness, behavior modification, and improved decision-making about solid waste disposal.

The involvement of the Department of Education is an important indicator of state commitment to a recycling education program. The Education Department is a good resource for networking with environmental teachers, offering professional input concerning curriculum development, assisting with distribution and teacher training, and providing funding and expertise to facilitate curriculum implementation. In most states, the Department of Education was at least minimally involved in recycling curriculum programs, often as a representative on an advisory committee. Recycling education programs, however, rarely originated in state Departments of Education. In some cases, the advisory role reflects the educational structure of the state, and characterizes the Department's involvement in all curriculum development. In general, however, lack of initiative and substantial commitment by the Department of Education reflects the delegation of responsibility for all aspects of solid waste management to that division of the Department of Environmental Protection.

Departments of Environmental Protection are universally involved in production of educational materials. In 17 of the 20 respondent states with programs, the DEP carried primary responsibility for education on all levels. In some cases, other departments within the environmental agency contributed to curriculum development, such as Water Quality or Energy Conservation. In eight states, other state departments contributed to materials development, depending upon the particular focus of the program.

The most critical characteristic of these programs concerns the extent of active cooperation between participating departments. A clear correlation exists between the broader, comprehensive educational programs and the degree of interdepartmental coordination in development, distribution and revision. In only five states was such cooperation an institutional part of the development procedure—California, Maine, Minnesota, Vermont, and Washington. In nine states, a moderate amount of constructive interaction characterized the program—Florida, Illinois, Indiana, Kentucky, Michigan, New Jersey, New York, Oregon, and Virginia. In some states the lack of interdepartmental communication and cooperation was prevalent enough to raise disturbing questions about the ability of state governments to pursue important environmental education goals, and effectively cope with solid waste management demands.

The Washington approach is the most permanent in its institutional nature. The Interagency Environmental Education Committee, created in 1985, has representatives from the Department of Ecology, Office of environmental Education (Department of Education) which work together to advise development and promotion of EE materials. The waste curriculum is one product of this system, and clearly draws upon an extensive financial and personnel base.

A commitment to structural communication and cooperation between agencies is prerequisite to constructive participation in a nationwide network such as the NNEE program. With a few exceptions (WA), state Education Departments have not assumed a leadership role in the development of recycling educational materials, although there is clear recognition of the need for such materials by Departments of Education.

Curriculum Development and Review

The curriculum developmental procedure provides necessary data for critical evaluation of curriculum materials; it reveals program priorities, and offers insight into relative program strengths and weaknesses. From a professional educational perspective, information concerning materials development is particularly important if the curricula are produced by Departments of Environmental Protection. Details concerning curriculum development are elaborated in the text, and summarized in Table 1.

Although the Departments of Environmental Protection generally oversee the production and distribution of educational materials, many agencies utilize environmental organizations or professional educational developers. Contracting with outside firms or individuals ensures that materials are completed within a specific time frame, and that projects are not endangered by staff or funding changes. Six respondent states worked with educational organizations or individuals—California, Kentucky, Michigan, Minnesota, Oregon, and Virginia—and eight worked with environmental organizations to develop materials—Maine, Michigan, New Jersey, Oregon, Rhode Island, Texas, Vermont, and Washington. State Universities are an invaluable resource, and Schools of Education are often affiliated with public schools, providing an opportunity to pilot and field test new materials. Very few states, however, actively utilized their university systems or schools of education to assist with curriculum development. Only six states involved universities, and none of them drew extensively upon the expertise and research resources available. One notable exception is the program coordinated by the Tennessee Valley Authority (TVA), technically a regional government body. The TVA has developed an exemplary environmental education program based in university centers throughout its service region. Each center conducts research, develops materials, offers training workshops, and serves as an effective link between communities. The TVA provides an overreaching administration to facilitate center cooperation, synthesize research, and provide initial funding. The Alliance for Environmental Education is using the TVA as a model for its National Network for Environmental Education program.

Analysis of development procedures reveals significant repetition, but also important similarities indicating that outside materials can be easily incorporated. Nearly all of the states or private groups began curriculum development with a search for existing materials. Although these searches varied in scope, they shared the purpose of selection and utilization of existing curricula and employed similar search techniques—databases, informal networking, bibliographies compiled by other individuals or states. Participation in the National Network for Environmental Education could provide agency staff with ready access to updated environmental educational materials and research documents, to streamline program development and avoid inefficient repetition.

After compilation of existing materials, nearly all states called upon volunteer teachers to develop, adapt, and evaluate classroom activities. State teachers were generally involved through multiple stages of development, pilot testing, revision, and final distribution. The scope of involvement varied tremendously, from five teachers employed for a summer to 40 teachers brought in for a 1-day workshop. These teachers usually offered their services without compensation, although most agencies paid travel expenses and some offered a small stipend.

Half of the respondent states performed some kind of preliminary implementation testing. The usual pilot program consisted of curriculum implementation by a group of volunteer teachers, who then evaluated the materials and made recommendations to the state agency. Curriculum materials include questionnaires to be completed and returned by future users for evaluation and revision purposes. Only California and Washington collected and analyzed specific field test, pilot program, or evaluation data on their education programs. States credited with "formal review structures" in Table 1 pursued a comprehensive revision strategy. The usefulness of data collection is contingent upon the necessary staff for data analysis, and subsequent resources to incorporate results into curriculum materials. Structured review procedures require longer-term resource allocation, which did not characterize most state attempts at curriculum development.

Teacher training workshops were almost universally offered, clearly considered the most effective way to encourage classroom implementation of materials. This training procedure, however, is both time consuming and labor-intensive for state agency staff. Many workshop programs functioned also to train teachers or recycling coordinators to teach their own series of workshops, thus accelerating the distribution process. Eighteen states offer workshops and encourage teacher attendance through a variety of communication and publicity channels, such as newsletters, press releases, and personal contact with school principals and superintendents. States pursue a variety of other strategies to promote teacher participation: four require attendance to receive a free curriculum copy—California, Kentucky, Rhode Island, and Washington; six offer in-service credit—Connecticut, Indiana, Kentucky, Missouri, Oregon, and Washington; Washington also pays substitute fees, and Virginia has developed video tapes to train new curriculum users.

Curriculum Description

State agencies have produced and distributed curriculum materials ranging from single magazine workbooks to comprehensive, sequential curriculum packages. The limited space in Table 1 prohibits the description of the specific structure of materials. In the text, each state program section contains a more detailed description of the curriculum materials, structure, goals, philosophy, and educational strategy. One half of the respondent states offers curriculum packages, which included teacher background guides, suggested student activities, student evaluations, and supplemental materials such as film strips, coloring books, posters, stickers, etc. The majority of states made supplemental materials available upon request to any interested individuals. Such optional materials require that teachers commit greater time and energy and are therefore less likely to be widely integrated into the classroom activities. Four curricula include direction on beginning school recycling programs—California, Massachusetts, Oregon and Washington.

To address the problem of time constraints in the classroom, all of the materials are flexible and voluntarily implemented. Some programs do have minimum time periods for classroom use, or suggested time frames, such as one concentrated week of activities. These time-specific programs are indicated in Table 1, but none have rigid time parameters. Since most curriculum developers drew upon teacher expertise, the materials are generally extremely "teacher friendly" and adaptable to local school district curriculum requirements.

All materials cover solid waste recycling topics, and many include hazardous waste and related water and air pollution issues. Solid waste management issues are inherently interdisciplinary and provide an excellent opportunity for integrated teaching. Almost all of the curriculum materials are deliberately multidisciplinary; a few are geared more toward the scientific or social scientific disciplines, especially in grades 9-12. Although many programs are designed to cover grades K-12, an emphasis on the upper elementary grades is evident. All of the programs include materials for the fourth through sixth grades, with the majority also offering activities for kindergarten through third grade, and the secondary materials focusing on the middle school grades. Half of the programs were validated by the respective state Departments of Education, although lack of validation can reflect the role of the Department in the state.

In addition to the issue of time constraints, there is the question of placement of environmental curricula materials within a comprehensive health education program. The California State Framework for Health Instruction, for example, includes ten health content areas that represent the scope of health instruction. These include:

Personal Health
Family Health
Nutrition
Mental-Emotional Health
Use and Misuse of Substances
Diseases and Disorders
Consumer Health
Accident Prevention and Emergency Services
Environmental Health
Community Health

Another example of a comprehensive education plan is provided by the Holyoke, Massachusetts public school system. In this example the content areas are represented in the various grade levels.

Elementary School Level—PK-5

Health, nutrition, safety awareness, environmental health, substance abuse, child abuse and human sexuality

Middle School Level—Grades 6-9

Health and human development, human sexuality, substance abuse and child abuse, nutrition, environmental health and exercise

High School Level
Human sexuality, teenage pregnancy, substance abuse, environmental health, nutrition and exercise

Teacher training, as has been mentioned above, is crucial for successful integration of any content area into a comprehensive health education curriculum. The Holyoke system is in the process of offering all teachers an opportunity to take course work in health areas including environmental health. This is in addition to required training for health teachers at all grade levels.

The educational quality of the materials can be partially assessed from the materials described, but depends ultimately upon the success of the program in reaching teachers and encouraging them to integrate materials effectively into their classroom activities. Materials assessment can be based upon development, testing, and review procedures. Program assessment necessarily includes an analysis of state financial and personnel commitment to communication, distribution, teacher training, follow-up contact, and revision. The most comprehensive curriculum programs demonstrate substantial and consistent commitment to all program components, and provide models and materials for outside states considering new or expanded programs.

Discussion and Recommendations

In the past few years, many states have begun recycling education programs as a response to the solid waste crisis. Development of easily integrated and technically accurate activities and teacher guides requires utilization of expertise in the fields of education and environmental management. Teacher organizations, state universities, environmental groups, and state government staff offer this expertise. At the same time, many materials are in use on the local level, due to individual teacher initiative, community environmental centers, or municipal recycling programs. A survey of existing materials in state classrooms can form the basis of an educational needs assessment, which is prerequisite to pursuing an efficient development program. Combining local materials with curriculum activities (and structure) borrowed from other states can create a program that is sufficiently state-specific in a short period of time.

Teachers, students, parents, and administrators should know about the curriculum during development, and provide as much input as possible. Field testing and pilot programs provide good opportunities to obtain input from the range of people involved in curriculum implementation. Communication is critical on all levels — between the federal and state governments, different state agencies, agencies and school districts, school districts and principals, principals and teachers, teachers and students, students and parents. A cooperative infrastructure of this nature, such as the Washington state interagency Environmental Education Committee, should characterize all environmental educational efforts. The committee develops and promotes environmental education materials, and can coordinate programs to accomplish broader curriculum goals.

Many recycling education programs pursue ambitious goals: to encourage students to reevaluate their understanding of waste and alter decision-making and ultimately behavior concerning solid waste disposal. Such goals assume a certain transfer of values from study of materials covered in the classroom to life outside of a school context. Any significant behavioral change or appreciably increased awareness depends upon strong reinforcement of curriculum materials and school recycling projects. The continuity of waste management education and experience is essential for societal and personal change. Effective curriculum programs are part of larger educational efforts, coordinated with significant changes in solid waste management policies in the state, such as construction of Materials Reprocessing Facilities (MRFs), implementation of curbside or other pick-up programs, and use of recycled materials by state governments. Environmental education only can be ultimately effective if it is supported by consistent environmental policy and reinforced in its broader social context.

It is important to expose children to environmental education throughout their public schooling, but the importance of recycling education must be evaluated relative to the many other responsibilities placed upon the public schools. Recycling and the solid waste crisis compete with sex and health education, drug awareness, and other urgent social issues for teacher commitment and incorporation into limited classroom time. Broad curriculum decisions generally fall to local school boards, or state Departments of Education, but teachers carry tremendous responsibility to coordinate the multitude of subjects and issues they must cover.

The premise of recycling education programs places a seemingly disproportionate social burden on public school teachers, and allocates to students a disproportionate amount of future responsibility for long-term solutions. As waste management costs consume portions of municipal budgets comparable to education expenditures, it seems appropriate to allocate resources for long-term waste management strategies to education. It is not reasonable, however, to expect waste management solutions to emerge from the state's public schools.

Considering the scope and magnitude of the solid waste crisis, it may also be unreasonable to place the management burden exclusively upon state Departments of Environmental Protection. The current crisis has emerged from a multitude of private and public sources; any comprehensive solution will require adjustment and commitment from all parties. Such commitment requires consistent funding and support. Many programs have stagnated after significant effort and resource investment due to shifting budget priorities. The expenditures are ultimately inevitable; allocations now can only help to allay future costs.

Index

Printed and bound by CPI Group (UK) Ltd, Croydon, CR0 4YY

17/10/2024

01775687-0013